Evolution Extended

Evolution Extended

Biological Debates on the Meaning of Life

edited by Connie Barlow

The MIT Press
Cambridge, Massachusetts
London, England

This book was set in Sabon by DEKR Corporation and was printed and bound in the United States of America.

Library of Congress Cataloging-in-Publication Data

Evolution extended : biological debates on the meaning of life /
  edited by Connie Barlow.
      p.   cm.
    Includes bibliographical references (p.   ) and index.
    ISBN 0-262-02373-3
    1. Evolution. 2. Evolution (Biology) 3. Evolution—Religious
aspects. 4. Evolution (Biology)—Religious aspects. 5. Evolution—
Religious aspects—Christianity. 6. Evolution (Biology)—Religious
aspects—Christianity. 7. Creationism. I. Barlow, Connie C.
B818.E838   1994
116—dc20   93-42025
                                                                        CIP

Contents

I know so little about an oyster's logic,
or why slugs mate acrobatically
from slime gallows.
Earth isn't small enough for me
to exhaust. Why covet mind-teasers
lightyears away?

—*Diane Ackerman*

The debates you will encounter in this book I learned none of while in school. Granted, some of the arguments were not as much as a gleam in anyone's eye two decades ago. But even then the classic works that form the core of this book—those by Julian Huxley, George Gaylord Simpson, Pierre Teilhard de Chardin, Jacques Monod, Charles Darwin—were plumping up library shelves. Classes in my major (zoology) helped me to know and to name but not particularly to reflect. Courses in Buddhism and metaphysics were interesting but too far removed from the wonders of nature that had always been my passion. I was a victim of the schism in what C. P. Snow has called "the two cultures."

So I read Hermann Hesse and Will Durant and a little Bertrand Russell on my own. A friend told me excitedly one day that our godlike professor of European intellectual history had recommended that he read Teilhard de Chardin, but I had never learned the skills of seeking counsel from the great and the powerful. Besides, college was ending and the grand adventure of life just beginning. There were too many mountains to climb and coves to kayak in my adopted home, Alaska, to bother much with books for the next decade or so. I did come upon some of the new writings that made physics popular with philosophical questers, but for me protons and photons were too abstract, too removed from the pulsing of life. The ubiquitous slugs of the southeast Alaska rainforest could teach me more.

Years later, when my professional phase of life as an energy economics consultant in Seattle was beginning to lose its luster, I chanced upon a lecture by James Lovelock on the Gaia hypothesis. While still under his spell, I visited a section of the bookstore I had never bothered with before. Not too far from my beloved "Nature" section, "Science" was a treasure trove. And the books there bore no resemblance to my useful but dreary college texts. I had discovered a world in which the two cultures were bridged.

Out of that discovery came this anthology series. *From Gaia to Selfish Genes: Selected Writings in the Life Sciences,* published in 1991, was a potpourri. Works of literary merit, these writings by biologists were as philosophically suggestive as they were scientifically rigorous. United by a propensity to suck us out of our organismic skins—to help us see things from the standpoint of the gene, the cell, living partnerships, and the planet—these writings spanned the gamut of research programs.

This second volume is unabashedly philosophical. Evolutionary biology is a powerful tool for honing one's worldview. It is rich for extension into the realm of meaning, grounding ultimate mysteries in foundations of facts. The inspiration here owes to several books that introduced me to the grand debate between Julian Huxley and George Gaylord Simpson on the question of evolutionary progress. The scientists were dead, but the ideas were smoldering with life. Soon I stumbled on two other books that also had the aura of great minds doing battle on a cosmic scale: *The Phenomenon of Man* by the Jesuit paleontologist Pierre Teilhard de Chardin and *Chance and Necessity* by the molecular biologist Jacques Monod, an existentialist—both also dead. I discovered other biologists, living and long gone, who likewise fused humanities and science in their writings on issues of vast philosophical consequence. You will meet those who have dealt with cosmologies here; in future volumes I hope to present the writings of those who have extended evolutionary biology into the realm of ethics and mused about the limits of life.

I have fashioned this collage of writings to give it the feel of a salon. The voices reappear, form alliances, and take issue with one another by name, revealing along the way the flesh-and-blood humans engaged in the drama of worldviews in collision. Scientists do not, of course, have *the* answers for questions of meaning. "After science has done its best," says John Burroughs, "the mystery is as great as ever, and the imagination and the emotions have just as free a field as before." Whatever humanistic, rationally derived, experiential, intuitive, mystical, or revealed knowledge each of us may consciously or unconsciously use in crafting a worldview, we owe it to ourselves to make an acquaintance with the best factual knowledge that the sciences can offer. In my view, there is no more pleasurable way to do so than in the context of the cosmic questions that biologists themselves have set about answering.

Those who wish will be able to make this journey without deserting cherished gods and demons; somebody surely will say what you most want to hear. Readers hungering for a spiritual overhaul can look forward to a number of jolts and surprises. Others may venture forth for the sheer fun of it; for them, the words of Ralph Waldo Emerson can serve as credo: "Were I to hold the truth in my hand I would let it go for the positive joy of seeking." For all, I hope the experience will build empathy for and tolerance of alien convictions. For me, assembling this book has been all these things and more.

Note to the Reader

All excerpts from previously published works appear in roman type, with authors and sources credited at the beginning of each. The original, contributed essays in chapter 8 also appear in roman. *Italics signifies my own editorial passages and insertions,* as well as standard usage of italic type within roman text. British spelling and punctuation in copyrighted works have been retained. To ensure a seamless presentation, deletions are not marked by ellipses. For short excerpts the bibliography records the page citations from the original. For long excerpts the concordance presents page citations, keyed to the headed sections and the paragraph order used in this anthology.

Chapter titles, section headings, and choice of visuals and poetry are my own doing. I deeply appreciate the flexibility of all living authors whose works are excerpted; they graciously consented to editorial decisions that compressed and sometimes rearranged their work.

1 Is Evolution Going Anywhere?

The scientific doctrine of progress is destined to replace not only the myth of progress, but all other myths of human earthly destiny. It will inevitably become one of the cornerstones of man's theology, or whatever may be the future substitute for theology, and the most important external support for human ethics.

—*Julian Huxley, 1957*

Progress is a noxious, culturally embedded, untestable, nonoperational, intractable idea that must be replaced if we wish to understand the patterns of history.

—*Stephen Jay Gould, 1988*

1 Prophets of Progress

Is evolution going anywhere? And where does humankind fit within the pageant of life? These questions grow out of our awareness of evolutionary biology, and a search for answers—personal answers—should be grounded in the science. But it is also clear that this kind of philosophical quest takes us beyond biology into the realm of meaning.

Julian Huxley has been one of the most influential advocates of a scientific vision of progress. Other biologists and philosophers schooled in biology have propounded progressive worldviews, but Huxley was one of the first to steer clear of nonmaterial causes. He was convinced that natural selection alone could do the job. The same process that Charles Darwin had identified in 1859 as the cause of evolution was, in Huxley's mind, infusing the history of life with a progressive glow.

Julian Huxley (1887–1975) was a British biologist who entered the field at a time when the original form of Darwinism had been eclipsed by the new-found and seemingly incompatible wonders of genetics. Darwinian natural selection and Mendelian genetics were finally reconciled in the 1930s and 1940s through the birth of the modern synthesis. That was Huxley's name for the achievement, and he was one of many biologists who helped bring it about. Along the way, Huxley was formulating a vision of evolutionary progress.

Julian Huxley was a prolific writer, uncommonly good at crafting technical books with popular appeal. In 1953 he was awarded the Kalinga Prize "for distinguished popular writing in science." His success as a popularizer did not, however, discredit his career as a scientist. Three years after the Kalinga award, Huxley received the Darwin medal of the Royal Society "for his distinguished contributions to the study and theory of evolution." In addition to his theoretical work in evolutionary biology, Huxley's field research on the courtship behavior of birds helped make the study of animal behavior a true science. His experimental work in developmental biology included the discovery that a certain species of salamander that had no air-breathing, terrestrial stage in its maturation could be made to lose its gills and seek land if it was fed the hormone-rich tissue of thyroid glands.

Julian Huxley's commitment to a progressive worldview was evident in many of his books and essays. Excerpts here (drawn from four of his books) present the biological arguments on which that worldview was

Progress: Myth or Science?

Julian Huxley

One of the myths of human destiny is that of progress. Thirty years ago, John B. Bury wrote a very interesting book on *The Idea of Progress (1920)*. When I read it, I was surprised at the modernity of the notion. Its history dates back to little more than three hundred years ago, and it is eminently a nineteenth-century concept. Other periods thought in terms of deterioration from a Golden Age, or of cyclical recurrence, or of the mere persistence of human sin and misery, tempered by hopes of salvation in another life.

The idea of progress could not have become part of general thought until men could see that, in one respect or another, they were improving their lot. In the eighteenth century, the chief emphasis seems to have been on the superiority of the civilized and cultured life of the period. In the nineteenth it was switched to the rapid improvement in man's technological control of nature; but there were many variants of the idea, ranging from the perfectability of man through universal education to the Hegelian doctrines about the National State. Darwin's work added scientific respectability to the general concept; but in practice it was used to justify any philosophy of progress in vogue, including the Prussian conception of progress through struggle and war.

In our Western world the myth of progress has now fallen on evil days. It was attacked by many writers on the grounds that the idea of progress cannot be reconciled with the retrogressions of Fascism and Nazism and the horrors of the recent war. Among its most recent assailants is my brother Aldous. *(Aldous Huxley wrote the futuristic novel* Brave New World.)

We all know the disillusionment that has set in within the brief space of half a hundred years. How the orderly mechanisms of nineteenth-century physics gave way to strange and sometimes non-rational concepts that no one but mathematicians could grasp; how the idea of relativity, and its somewhat illegitimate extension into human affairs, destroyed faith in the absolute, whether absolute truth or absolute morality or absolute beauty; how our belief in the essential rationality and goodness of man was undermined by psychology and sent crashing in ruins by the organized cruelty of Belsen and the mass folly of two world wars; and how our idealistic notions of progress as the inevitable result of science and education were shattered by events. In brief, man's first evolutionary picture of nature and his own place in it proved false in its design and had to be scrapped.

Meanwhile, however, the patient labours of the students of evolution, whether stellar evolution, biological evolution, or social evolution, have revealed that progress is not myth but science, not an erroneous wish-fulfilment, but a fact. On the other hand, progress as a scientific doctrine reveals itself as very different from progress as a mythical dogma. The scientific doctrine of progress is destined to replace not only the myth of progress, but all other myths of human earthly destiny. It will inevitably become one of the cornerstones of man's theology, or whatever may be the future substitute for theology, and the most important external support for human ethics. There has not yet been time to work it out in detail; indeed, a number of facts relevant to its elaboration still await discovery. But its broad lines are now clear.

Critics of "Higher" and "Lower"

For millions and millions of years living substance was confined within the prison walls of microscopic floating cells. Who would have ventured to prophesy what it could bring forth? The flowers carpeting the soil, the great trees with the singing birds in their branches, the glistening fish among the reefs of coral, tribes of busy insects, the strength of the bull and the beauty of the butterfly, the elaborate fixed instincts of the bee, the intelligence and flexibility of behavior of the dog: these are among the wonderful and admirable possibilities that have been realized.

If we accept the doctrine of evolution, we are bound to believe that man has arisen from mammals, terrestrial from aquatic forms, vertebrates from invertebrates, multicellular from unicellular, and in general the larger and the more complex from the smaller and simpler. To the average man it will appear indisputable that a man is *higher* than a worm or a polyp, an insect is *higher* than a protozoan, even if he cannot exactly define in what resides this highness or lowness of organic type. It is, curiously enough, among the professional biologists that objectors to the notion of biological progress and to its corollary, the distinction of higher and lower forms of life, have chiefly been found.

The objections that have been made to employing *progress* at all as a biological term, and to the use of its correlates *higher* and *lower* as applied to groups of organisms, are as follows. First, it is objected that a bacillus, a jellyfish, or a tapeworm is as well adapted to its environment as a bird, an ant, or a man, and that therefore it is incorrect to speak of the latter as higher than the former, and illogical to speak of the processes leading to their production as involving progress. An even simpler objection is to use mere survival as criterion of biological value, instead

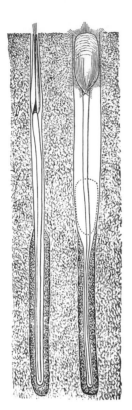

Lingula lives in ocean sediments and filters small organisms from currents at the mouth of its burrow. The dotted line indicates the position of the body when retracted.

of adaptation. Man survives: but so does the tubercle bacillus. So why call man the higher organism of the two?

A second class of objector is prepared to admit that there has been an increase of complexity, an increase in the degree of organization during evolution, but refuses to allow that increase of complexity has any value in itself, whether biological or philosophical, and accordingly refuses to dignify this trend towards greater complexity by the name of progress. Yet a third difficulty is raised by those who ask us to fix our attention on forms of life like Lingula, the lamp-shell, which, though millions of years elapse, do not evolve. If there exists a Law of Progress, they say, how is it that such creatures are exempt from its operations?

Finally, a somewhat similar attitude is adopted by those who refuse to grant that evolution can involve progress when it has, as we know, brought about well-nigh innumerable degeneration. Degeneration is a form of specialization in which the majority of the somatic organs are sacrificed for greater efficiency in adaptation to a sedentary or parasitic life. Locomotor organs disappear, sensory and nervous systems are much reduced, and in parasites the digestive system may be abolished. Reproductive mechanisms, however, may be inordinately specialized. The degradation of parasites and sedentary types is equally a product of the evolutionary process with the genesis of the ant, the bird or the human being: how then can we call the evolutionary process progressive?

These are important objections. Can they be met?

Countering the Critics

We can see why lower and higher forms can survive side by side. There is an enormous range of evolutionary possibility open to life, and each plan of construction can take advantage only of a certain fraction of it. The higher forms are those which deployed later in time, to take advantage of possibilities that earlier forms had not succeeded in realizing. But they are no more equipped to lead the life of the lower types than the lower types are equipped to lead their life. Mammals cannot take the place of worms, or worms fill the niche occupied by protozoa.

During evolution, the onward-flowing stream of life breaks up into a vast number of branches or trends, each resulting in improvement of one sort or another. The great majority of these become so specialized that life in them finds itself in a blind alley, incapable of further improvement or of transformation for another way of existence. After this, they either remain essentially unchanged for tens or even hundreds of millions of years, or else wholly die out, becoming extinguished in the sands of time.

Directional change is thus normally succeeded by stabilization. This obviously holds good for the relatively minor trends we call specializations, like the specialization of whales for a secondarily aquatic life or of horses for grazing and rapid running. But, as my grandfather T. H. Huxley was one of the first to point out, it holds good also for major organizational trends and improvements. The occupants of the biological scene are for the most part what he called "persistent types", which have remained unchanged in their essential characteristics from the moment when they have come up against their invisible limitations.

The most spectacular are the so-called "living fossils" like the lungfishes, which have persisted as rare survivors of a once abundant group for over three hundred million years; or the lampshell Lingula, which is barely to be distinguished from its ancestors preserved in the Ordovician rocks of four hundred million years ago. But an entire successful group may persist. The Coelenterates, such as jellyfish, polyps, and corals, certainly first became abundant and successful well before the beginning of the fossil record in the Cambrian, that is to say much more than five hundred million years ago; and they are still abundant and successful today. This does not mean that there has been no evolutionary change within the group during this portentous length of time. New specialized sub-groups, for instance of corals, have arisen and old sub-groups have died out. But the coelenterate level of organization has never been transcended by such new sub-groups: the coelenterate type of construction and working has persisted, though variations have been played on its essential theme.

The same holds true even for the latest and most finished products of evolution. The ants, in many ways the highest invertebrate type, have shown neither advance nor essential change since the time, some fifty million years ago, when ancestral specimens were trapped in the resin that has hardened into Baltic amber. The bird type has not changed in its basic quality of warm-blooded flying machine for perhaps twenty-five million years, although much minor specialization has taken place, adapting different avian lines to different habitats, niches, and ways of life.

Finally, to deny progress because of degeneration is really no more legitimate than to assert that, because each wave runs back after it has broken, therefore the tide can never rise. Similarly with the first two objections. If the degree of adaptation has not increased during evolution, then it is clear that progress does not consist in increase in adaptation. But it does not follow that progress does not exist; it may quite well consist in an increase of other qualities. So with complexity. Complexity has increased, but increase in complexity is not progress, say the objectors. Granted: but may there not be something else which has increased besides mere complexity?

Succession of Dominant Types

The task before the biologist is not to define progress *a priori*, but to proceed inductively to see whether he can or cannot find evidence of a process which can legitimately be called progressive.

Even the hardened opponents of the idea of biological progress find it difficult to avoid speaking of higher and lower organisms, though they may salve their consciences by putting the words between inverted commas. The unprejudiced observer will accordingly begin by examining various types of "so-called higher" organisms and trying to discover what characters they possess in common by which they differ from "lower" organisms. He will then proceed to examine the course of evolution as recorded in fossils and deduced from indirect evidence, to see what the main types of evolutionary change have been; whether some of them have consistently led to the development of characters diagnostic of "higher" forms; which types of change have been most successful in producing new groups, dominant forms, and so forth. If evolutionary progress exists, he will by this means discover its factual basis, and this will enable him to give an objective definition.

Proceeding on these lines, we can immediately rule out certain characters of organisms and their evolution from any definition of biological progress. Adaptation and survival, for instance, are universal, and are found just as much in "lower" as in "higher" forms: indeed, many higher types have become extinct while lower ones have survived. Complexity of organization or of life-cycle cannot be ruled out so simply. High types *are* on the whole more complex than low. But many obviously low organisms exhibit remarkable complexities, and, what is more cogent, many very complex types have become extinct or have speedily come to an evolutionary dead end.

Perhaps the most salient fact in the evolutionary history of life is the succession of what the paleontologist calls dominant types. These are characterized not only by a high degree of complexity for the epoch in which they lived, but by a capacity for branching out into a multiplicity of forms. This radiation seems always to be accompanied by the partial or even total extinction of competing main types, and doubtless the one fact is in large part directly correlated with the other.

In the early Paleozoic the primitive relatives of the Crustacea known as trilobites were the dominant group. These were succeeded by the marine arachnoids called sea-scorpions or eurypterids, and they in turn by the armoured but jawless vertebrates, the ostracoderms, more closely related to lampreys than to true fish. The fish, however, were not far behind, and soon became the dominant group. Meanwhile, groups both from among the arthropods and the vertebrates became adapted to land life, and towards the close of the Paleozoic, insects and amphibians could both claim the title of dominant groups. The amphibia shortly gave rise to the reptiles, much more fully adapted to land life, and the primitive early insects produced higher types, such as beetles, hymenoptera *(ants and wasps)* and lepidoptera *(butterflies and moths)*. Higher insects and reptiles were the dominant land groups in the Mesozoic, while among

CENOZOIC	0
Quaternary	
Tertiary	2
	65
MESOZOIC	
Cretaceous	
Jurassic	
Triassic	
	245
PALEOZOIC	
Permian	
Carboniferous	
Devonian	
Silurian	
Ordovician	
Cambrian	
	570
PROTEROZOIC	
	2500
ARCHEAN	
	3800

in millions of years BP

aquatic forms the fish remained pre-eminent, and evolved into more efficient types: from the end of the Mesozoic onwards, however, they show little further change.

Birds and mammals began their career in the Mesozoic, but only became dominant in the Cenozoic. The mammals continued their evolution through the whole of this epoch, while the insects reached a standstill soon after its beginning. Finally man's ancestral stock diverged, probably towards the middle of the Cenozoic, but did not become dominant until the latter part of the Ice Age.

Efficiency, Control, and Independence

If we then try to analyse the matter still further by examining the characters which distinguish dominant from non-dominant and earlier from later dominant groups, we shall find, first of all, efficiency in such matters as speed and the application of force to overcome physical limitations. The eurypterids must have been better swimmers than the trilobites; the fish, with their muscular tails, much better than either; and the later fish are clearly more efficient aquatic mechanisms than the earlier. Similarly the earlier reptiles were heavy and clumsy, and quite incapable of swift running. Sense-organs also are improved, and brains enlarged. In the latest stages the power of manipulation is evolved. Through a combination of these various factors man is able to deal with his environment in a greater variety of ways, and to apply greater forces to its control, than any other organism.

Another set of characteristics concerns the internal environment. Lower marine organisms have blood or body-fluids identical in saline concentrations with that of the seawater in which they live; and if the composition of their fluid environment is changed, that of their blood changes correspondingly. The higher fish, on the other hand, have the capacity of keeping their internal environment chemically almost constant. Birds and mammals have gone a step further: they can keep the temperature of their internal environment constant too, and so are independent of a wide range of external temperature change.

The early land animals were faced with the problem of becoming independent of changes in the moisture-content of the air. This was accomplished only very partially by amphibia, but fully by adult reptiles and insects through development of a hard impermeable covering. The freeing of the young vertebrate from dependence on water was more difficult. The great majority of amphibians are still aquatic for the earlier part of their existence.

There is no need to multiply examples. The distinguishing characteristics of dominant groups all fall into one or other of two types—those making for greater control over the environment, and those making for

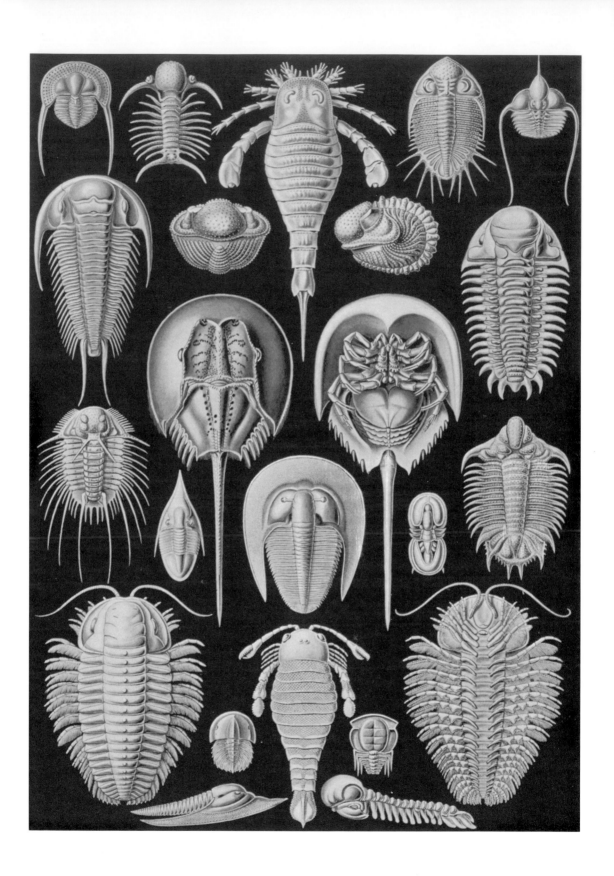

Trilobites (along borders) and eurypterids (top and bottom center) are extinct. The horseshoe crab (center; dorsal and ventral views) is a "living fossil."

greater independence of the environment. Thus advance in these respects may provisionally be taken as the criterion of biological progress.

It is important to realize that progress, as thus defined, is not the same as specialization. Specialization is an improvement in efficiency of adaptation for a particular mode of life: progress is an improvement in efficiency of living in general. The latter is an all-round, the former a one-sided advance. We must also remember that in evolutionary history we can and must judge by final results. And there is no certain case on record of a line showing a high degree of specialization giving rise to a new type. All new types which themselves are capable of adaptive radiation seem to have been produced by relatively unspecialized ancestral lines.

Looked at from a slightly different angle, we may say that progress must in part at least be defined on the basis of final results. These results have consisted in the historical fact of a succession of dominant groups. And the chief characteristic which analysis reveals as having contributed to the rise of any one of these groups is an improvement that is not one-sided but all-round and basic. Temperature-regulation, for instance, is a property which affects almost every function as well as enabling its possessors to extend their activities in time and their range in space. Placental reproduction is not only a greater protection for the young—a placental mother, however hard-pressed, cannot abandon her unborn embryo—but this additional protection, together with the later period of maternal care, makes possible the extension of the plastic period of learning which then served as the basis for the further continuance of progress.

It might, however, be held that biological inventions such as the lung and shelled egg, which opened the world of land to the vertebrates, are after all nothing but specializations. Are they not of the same nature as the wing which unlocked the kingdom of the air to the birds, or even to the degenerations and peculiar physiological changes which made it possible for parasites to enter upon that hitherto inaccessible habitat provided by the intestine of other animals? This is in one sense true; but in another it is untrue. The bird and the tapeworm, although they did conquer a new section of the environment, in so doing were as a matter of actual fact cut off from further progress. Theirs was only a specialization, though a large and notable one. Birds were ruled out by their depriving themselves of potential hands in favour of actual wings, and perhaps also by the restriction of their size made necessary in the interests of flight.

The conquest of the land, however, not only did not involve any such limitations, but made demands upon the organism which could be and in some groups were met by further changes of a definitely progressive nature. Temperature-regulation, for instance, could never have arisen through natural selection except in an environment with rapidly changing temperatures: in the less changeable waters of the sea the premium upon it would not be high enough. Lungs were one needful precursor of intelligence. Warm blood was another, since only with a constant internal environment could the brain achieve stability and regularity for its finer functions.

The Pageant of Life

There has been, we find, a succession of organizational types, the later-appearing ones possessing a higher level of organization than the earlier. Structural organization rises from the pre-cellular to the cellular and the multi-cellular level; there follows the multi-tissued type, like the sea-anemone, and then the multi-organed type like the worm or the mollusc or the early arthropod. The multi-organed animal attains new mechanical and physiological levels, as in crustacea and fish, new and superior modes of organization or reproduction, as in insects and reptiles, and new levels of behaviour appear, as in birds and mammals and social insects. New methods of integration and homeostatic adjustment arise, such as the endocrine system and the temperature-regulating mechanisms of higher vertebrates.

There has been an enormous rise in level of complex but harmonious organization—think of a bird or a mammal as against a flatworm or a jellyfish; in flexibility and the capacity for self-regulation; in physiological efficiency, as shown in muscular contraction or rate of nervous conduction, or manifested in sheer strength or speed; in the range of awareness, as seen in the evolution of sense-organs—think of an eagle's eyes or an antelope's ears as against the blindness and deafness of a polyp or amoeba; and in the intensity of complexity of mental processes such as knowing and perceiving, feeling and willing, learning and remembering— think of dogs or elephants as against sea-anemones or snails. If we speak of a cunning wolf or a wary crow, we imply that their life has taught them new qualities; but it is nonsense to talk of a cunning crab, and, though we might properly ascribe wariness to a trout, I would not like to speak of a wary Amoeba. In the same way we can justifiably say that one dog is affectionate, another intelligent: but to speak of an affectionate earthworm or an intelligent snail has no more proper significance than it would be to say that a dog was intellectual or religious.

An apparently progressive advance may turn out ultimately to be a limitation. For instance, the insects successfully conquered the land by developing a method for breathing with the aid of fine air-tubes penetrating every tissue of the body. This constitutes an admirable mechanism so long as the creature remains small, but makes large size impossible. An insect as big as a rat just wouldn't work properly: actually no insect is bigger than a mouse. This limitation of total size naturally sets a limit to the size of the brain, and so to the number of cells in the brain, which in turn sets a low limit to the degree of intelligence and the flexibility of behaviour. That is why insects are never very intelligent, but have to

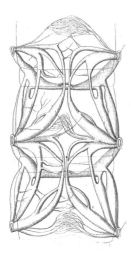

Anatomy of air supply system of a typical millipede or centipede. Section shows three pairs of spiracles, the holes through which air enters and leaves the system, connected with a network of tubes.

depend mainly on the often marvellous but always rigid and limited behaviour-mechanisms we call instincts. This limitation of insect size is very lucky for us, because without it, man assuredly could never have evolved.

I must bring this section to a close by reminding you of the greatest improvement ever made in the machinery of life—the improvement of the nervous system. Before the nervous system could be improved, it had to be invented: the lowest animals are without one. Even the largest sponge has no nervous system whatever. Its first manifestation, which we find in creatures like polyps, is a nerve net—an irregular network of nerve cells and interlacing fibres extending all over the body. Its latest improvement is our own nervous system, with all its incoming and outgoing wires of nerves, and its central exchange and office and control room, in the form of an enormous brain filling up most of our head. During evolution, the speed at which messages are transmitted along nerve fibers has increased over six hundredfold, from below six inches a second in some nerve nets to over a hundred yards a second in parts of our own nervous system. The brain's complexity of organization is almost infinitely greater than that of any other piece of biological machinery in ourselves or in any other animal.

In mammals and birds, each generation can extend its influence on to the next, and the experience of the parents is in part made available to the offspring. But never until the origin of speech was it possible for a whole series of generations to be linked together by experience, never could experience be cumulative, never could one mind know what another mind, remote in time, had been thinking or feeling.

Defining Evolutionary Progress

During the course of evolution in time, there has been an increase in the control exerted by organisms over their environment, and in their independence with regard to it; there has been an increase in the harmony of the parts of organisms; and there has been an increase in the psychical powers of organisms, an increase of willing, of feeling, and of knowing. This increase has not been universal; many organisms have remained stationary or have even regressed; many have shown increase in one particular but not in others. But the *upper level* of these properties of living matter has been continually raised, their average has continually increased. It is to this increase in the average and especially in the upper level of these properties that, I venture to think, the term biological progress can be properly applied. We have thus arrived at a definition of evolutionary progress as consisting in a raising of the upper level of biological efficiency, this being defined as increased control over and independence of the environment.

*Duck on Rock in
Snowy Moonlit Night*
(Shibata Zeshin)

I personally would like to see a new evolutionary classification, which would combine the advance and ancestry principles. We would have groups (or "clades", from the Greek for branches) of common ancestry—classes, order, and other familiar designations—and grades of advance for which new designations would be needed. Thus, Birds and Mammals would continue to rank as two classes, but would be included in a single grade, which might be called *Homotherma,* since temperature-regulation is their diagnostic improvement. Other obvious grade labels for Vertebrates would include terms already in use, such as *Gnathostomata* for forms with jaws, *Tetrapoda* for those with walking limbs, and *Amniota* for those with a protective "private pond" for the embryo. I would hope that *Metazoa* would be restored to its original use as a grade label [for multicellular animals with specialized tissues coordinated by a nervous system] and that Man would be placed in a new major grade, which might be called *Psychozoa.*

Is Progress Inevitable?

I want now to deal with the question of the inevitability of progress. In biological evolution progress is in one sense incvitable, in another sense not. It is inevitable in the sense that, given the struggle for existence and natural selection in our world or any world similar to ours during the last thousand million years, it is apparently unavoidable that true progress should occur in some of the lines of life. But it is not universally inevitable: the great majority of biological stocks either show no progress, the reverse of progress, or a progress which is only partial and limited. It is conditioned by accidents; if the identical stock which showed progressive evolution on a continent could have been transplanted to a small oceanic island with different competitors, it would assuredly not have progressed. If the world had not had the accident of a great climatic catastrophe befall it at the close of the Cretaceous, the ancestral mammals would not have supplanted the reptiles so completely nor embarked upon such rapid new advance. And it will always remain subject to accidents. If some virus or bacterium were to arise which exterminated the human species, that would almost certainly be the end of any hopes of major progress on earth.

Progress is inevitable as a general fact; but it is unpredictable in its particulars. The changes which would confer advantage in the struggle for existence may take place in any direction—with, or against, or at right angles to the stream of progress. By means of those which march with that stream, the upper level of life's attainment is raised. But the struggle still goes on: and again, starting from this new condition, there will be variations in every direction which will have survival value, and some of these will be progressive; and so the upper level will be once more raised. There will further be a premium upon progressive changes, since a progressive change will generally land its possessor in virgin soil, so to speak; if not in an actually new physical environment, then in a biologically new situation.

Accounting for Cruelty

Thus from the well-established biological premises of (1) the tendency to geometrical increase with consequent struggle for existence and (2) some form of inherited variability, we can deduce as necessary consequence, not only the familiar but none the less fundamental fact of Natural Selection, but also the almost neglected fact that a *certain fraction* of the guiding force of Natural Selection will inevitably be pushing organisms into changes that are progressive. What is more, there was progress before man ever appeared on earth, and its reality would have been in no way impaired even if he had never come into being. His rise only continued, modified, and accelerated a process that had been in operation since the dawn of life.

Here we find, in the intellectual sphere at least, that assurance which men have been seeking from the first. We see revealed, in the fact of evolutionary progress, that the forces of nature conspire together to produce results which have value in our eyes, that man has no right to feel helpless or without support in a cold and meaningless cosmos, to believe that he must face and fight forces which are definitively hostile.

There is, of course, a problem of values lurking in the background. How, for instance, can we reconcile the existence of so much cruelty and suffering with the concept of progress? In general, parasites have evolved at the expense of their hosts, often causing great suffering, as with the fly maggots that live in the noses of various animals. Furthermore, the adaptations of parasites often involve large-scale degeneration in the parasites themselves, especially of locomotor and sensory organs. How can this be reconciled with the notion of advance?

There is no simple answer. The specializations of a liver fluke for a parasitic existence are improvements from the angle of the parasites' own evolution. Pain and suffering are part of the wastage involved in the workings of the selective process. We must not expect to find human values at work in nature's day-to-day operations.

What is remarkable, it seems to me, is that the blind and automatic forces of mutation and selection, operating through competition and focused immediately on mere survival, should have resulted in anything that merits the name of advance or progress. When looked at on the large scale, biological evolution *has* resulted—not universally, but regularly—in the overcoming of limitations, and has led to a steady rise in the upper level of life's achievements. It has produced co-operation as well as competition, and it has led finally to the emergence of values as operative factors in the process. These are facts to be accepted, not theories to be reconciled with other theories.

Excesses of Anthropomorphism

This brings us to a further objection which is often raised to the idea of progress, namely that it is a mere anthropomorphism. This view asserts that we judge animals as higher or lower by their greater or lesser resemblance to ourselves and that we give the name progress to the evolutionary trend which happens to have culminated in ourselves. If we were ants, the objectors continue, we should regard insects as the highest group and resemblance to ants as the essential basis of a "high" organism: while if we were eagles our criterion of progress would be an avian one.

J. B. S. Haldane *(a prominent evolutionary biologist)* has adopted this view. He writes:

I have been using such words as 'progress', 'advance', and 'degeneration', as I think one must in such a discussion, but I am well aware that such

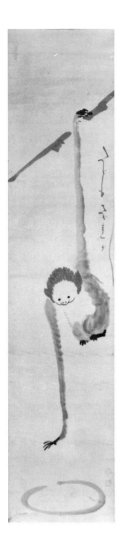

Monkey Reaching for the Moon in the Water (Hakuin Ekaku)

terminology represents rather a tendency of man to pat himself on the back than any clear scientific thinking. . . . Man of today is probably an extremely primitive and imperfect type of rational being. He is a worse animal than the monkey. . . . We must remember that when we speak of progress in Evolution we are already leaving the relatively firm ground of scientific objectivity for the shifting morass of human values.

This I would deny. Haldane has neglected to observe that man possesses greater power of control over nature, and lives in greater independence of his environment than any monkey. The use of an inductive method of approach removes all force from such objections. The definitions of progress that we were able to name as a result of a survey of evolutionary facts, though admittedly very general, are not subjective but objective in their character.

That the idea of progress is not an anthropomorphism can immediately be seen if we consider what views would be taken by a philosophic tapeworm or jellyfish. Granted that such organisms could reason, they would have to admit that they were neither dominant types, nor endowed with any potentiality of further advance, but that one was a degenerate blind alley, the other a specialization of a primitive type long left behind by more successful forms of life. And the same would be equally true, though not so strikingly obvious, of ant or eagle. Man *is* the latest dominant type to be evolved, and this being so, we are justified in calling the trends which have led to his development *progressive.* We must, however, of course beware of subjectivism and of reading human values into earlier stages of evolutionary progress. Human values are doubtless essential criteria for the steps of any future progress: but only biological values can have been operative before man appeared.

Human Purpose in Evolution

The ordinary man, or at least the ordinary poet, philosopher, and theologian, is always asking himself what is the purpose of human life, and is anxious to discover some extraneous purpose to which he and humanity may conform. Some find such a purpose exhibited directly in revealed religion; others think that they can uncover it from the facts of nature. One of the commonest methods of this form of natural religion is to point to evolution as manifesting such a purpose. The history of life, it is asserted, manifests guidance on the part of some external power; and the usual deduction is that we can safely trust that same power for further guidance in the future.

I believe this reasoning is wholly false. The purpose manifested in evolution, whether in adaptation, specialization, or biological progress, is only an apparent purpose. It is just as much a product of blind forces as is the falling of a stone to earth or the ebb and flow of the tides. It is we who have read purpose into evolution, as earlier men projected will and emotion into inorganic phenomena like storm or earthquake. If we wish to work towards a purpose for the future of man, we must formulate that purpose ourselves. Purposes in life are made, not found.

But if we cannot discover a purpose in evolution, we can discern a direction—the line of evolutionary progress. And this past direction can serve as a guide in formulating our purpose for the future. Increase of control, increase of independence, increase of internal coordination; increase of knowledge, of means for coordinating knowledge, of elaborateness and intensity of feeling—those are trends of the most general order. If we do not continue them in the future, we cannot hope that we are in the main line of evolutionary progress any more than could a sea-urchin or a tapeworm.

The precise formulation of human purpose cannot be decided on the basis of the past. Each step in evolutionary progress has brought new problems, which have had to be solved on their own merits. With the new predominance of mind that has come with man, life finds its new problems even more unfamiliar than usual. This last step marks a critical point in evolution, and has brought life into situations that differ in quality from those to which it was earlier accustomed.

Human purpose and the progress based upon it must accordingly be formulated in terms of human values; but it must also take account of human needs and limitations, whether these be of a biological order, such as our dietary requirements or our mode of reproduction, or of a human order, such as our intellectual limitations or our inevitable subjection to emotional conflict.

Obviously the formulation of an agreed purpose for man as a whole will not be easy. There have been many attempts already. Today *(1943)* we are experiencing the struggle between two opposed ideals—that of the subordination of the individual to the community, and that of his intrinsic superiority. Another struggle still in progress is between the idea of a purpose directed to a future life in a supernatural world, and one directed to progress in this existing world. Until such major conflicts are resolved, humanity can have no single major purpose, and progress can be but fitful and slow. Before progress can begin to be rapid, man must cease being afraid of his uniqueness, and must not continue to put off the responsibilities that are really his on to the shoulders of mythical gods or metaphysical absolutes.

But let us not forget that it is possible for progress to be achieved. After the disillusionment of the early twentieth century it has become as fashionable to deny the existence of progress and to brand the idea of it as a human illusion, as it was fashionable in the optimism of the nineteenth century to proclaim not only its existence but its inevitability. The truth is between the two extremes. Progress is a major fact of past evolution; but it is limited to a few selected stocks. It may continue in the future, but it is not inevitable; man, by now become the trustee of evolution, must work and plan if he is to achieve further progress for himself and so for life.

Human Society in the Cosmic Process

In the human phase of evolution, the struggle for existence has been largely superseded, as an operative force, by the struggle for fulfillment. It is the combination of these two terms which seems to me important. Human life *is* a struggle—against frustration, ignorance, suffering, evil, the maddening inertia of things in general; but it is also a struggle *for* something, and for something which our experience tells us can be achieved in some measure, even if we personally find ourselves debarred from any measure that seems just or reasonable. And fulfillment seems to describe better than any other single word the positive side of human development and human evolution—the realization of inherent capacities by the individual and of new possibilities by the race; the satisfaction of needs, spiritual as well as material; the emergence of new qualities of experience to be enjoyed; the building of personalities. But it cannot be achieved without struggle, not merely struggle with external obstacles, but with the enemies within our own selves.

The individual has duties not only to society but to himself—or perhaps I should say to the possibilities that are in him. It is true that one aspect of his fulfillment lies in working for others; but another aspect consists in his enjoyments and the free exercise of his capacities. For, after all, it is only through human individuals that the evolutionary process reaches its highest and most varied achievements. Each time you enjoy a sunset or a symphony, each time you understand an interesting fact or idea, each time you find satisfaction in making something, or in disciplined activity like sport, evolution has brought another of its possibilities to fruition.

In the light of evolutionary biology man can now see himself as the sole agent of further evolutionary advance on this planet, and one of the few possible instruments of progress in the universe at large. He finds himself in the unexpected position of business manager for the cosmic process of evolution. He no longer ought to feel separated from the rest of nature, for he is part of it—that part which has become conscious, capable of love and understanding and aspiration. He need no longer regard himself as insignificant in relation to the cosmos. He is intensely significant. In his person, he has acquired meaning, for he is constantly creating new meanings. Human society generates new mental and spiritual agencies, and sets them to work in the cosmic process: it controls matter by means of mind.

Biology has thus revealed man's place in nature. He is the highest form of life produced by the evolutionary process on this planet, the latest dominant type, and the only organism capable of further major advance or progress. His destiny is to realize new possibilities for the whole terrestrial sector of the cosmic process, to be the instrument of further evolutionary progress on this planet.

Through the doctrine of progress we can be both consoled and exhorted to effort; we can be guided and we can be warned; we can be given an enduring foundation, and also a goal. Our acceptance of the fact of progress and our understanding of the doctrine of progress constitute the major prerequisite for further progress.

———

Before I was born out of my mother, generations guided me;
My embryo has never been torpid—nothing could overlay it.

For it the nebula cohered to an orb,
The long slow strata piled to rest it on,
Vast vegetables gave it sustenance,
Monstrous sauroids transported it in their mouths, and deposited it
    with care.

All forces have been steadily employ'd to complete and delight me;
Now on this spot I stand with my robust Soul.

—*Walt Whitman*

———

An Alternative View of Progress

In 1974, the year before Julian Huxley's death, Francisco J. Ayala, a population biologist and a Spanish-born American, tried to salvage the notion of biological progress from what he believed were fatal flaws in Huxley's argument. First, Ayala pointed out that "progress" necessarily implies a value judgment; it cannot be defined as a biological term.

Francisco J. Ayala   The numerous writers on evolutionary progress have usually proceeded by identifying one or another attribute as the criterion of progress. A common deficiency in these discussions is the stated or implicit conviction that *the* criterion of progress has been discovered, often accompanied by a lack of awareness that progress is a value-laden concept rather than a strictly scientific one.

Once this subjectivity is recognized, however, Ayala sees that it is possible and perhaps useful to speak of evolutionary progress as directional change that, in particular contexts, may be regarded as an "advance." Directional change can be assessed scientifically. For example: Is a lineage increasing or decreasing in body size or population through time? Is it radiating into new niches? Is it losing digits? Is it putting more resources into attracting pollinators or nurturing its young? Such questions can range from the trivial to the profound, and scientists can look for evidence (or lack of evidence) of persistent directional change, casting their net over a narrow or wide group of taxa. But conflation of such directional change with the notion of "advance" or "progress" is a subjective act. Ayala is aware of the deep pull of the subjective; rather than decry such extensions, he suggests that scientists make an effort to manage them in ways that acknowledge the subjectivity while aiming for scientific goals of precision and utility.

If the term 'progress' were to be completely obliterated from scientific discourse, I would be quite pleased; but it seems to me unlikely to happen. The notion of progress seems to be irrevocably ingrained among the thinking categories of modern man and, hence, likely to continue being used in biology, particularly in reference to the evolutionary process. I have, therefore, attempted to clarify the concept in order to demythologize it.

Specifically, Ayala argues that by defining progress as "increased control over and independence of the environment," Huxley misread a kind of broad directional change that Ayala himself sees as profound, and hence worthy of attention. Rather, "the ability to perceive the environment, and to integrate, coordinate, and react flexibly to what is perceived" is Ayala's candidate for the most subjectively tantalizing directional change evident in the history of life.

Francisco Ayala's ideas appeared initially in a 1974 essay, "The Concept of Biological Progress," in Studies in the Philosophy of Biology, *edited by Ayala and Theodosius Dobzhansky. The slightly modified version here is drawn from Ayala's essay, "Can 'Progress' Be Defined as a Biological Concept?" in* Evolutionary Progress, *edited by Matthew H. Nitecki, copyright 1988 University of Chicago Press. It is reprinted here with permission of the author and the publisher.*

Increased ability to gather and process information about the environment is sometimes expressed as evolution towards "independence from the environment." This latter expression is misleading. No organism can be truly independent of the environment. The evolutionary sequence fish to amphibian to reptile allegedly provides an example of evolution toward independence from an aqueous environment. Reptiles, birds, and mammals are indeed free of the need for water as an external living medium, but their lives depend on the conditions of the land. They have not become independent of the environment, but have rather exchanged dependence of one environment for dependence on another.

The notion of "control over the environment" also has been associated with the ability to gather and use information about the state of the environment. However, true control over the environment occurs to any substantial extent only in the human species. All organisms interact with the environment, but they do not control it. Burrowing a hole in the ground or building a nest in a tree, like the construction of a beehive or a beaver dam, does not represent control over the environment except in a trivial sense.

The ability to obtain and process information about the conditions of the environment does not provide control over the environment but rather it enables the organisms to avoid unsuitable environments and to seek suitable ones. It has developed in many organisms because it is a useful adaptation.

Some selective interaction with the environment occurs in all organisms. The cell membrane of a bacterium permits certain molecules but not others to enter the cell. Selective molecular exchange occurs also in the inorganic world; but this can hardly be called a form of information processing. The most rudimentary ability to gather and process information about the environment may be found in certain single-celled eukaryotes (organisms with a true nucleus). A *Paramecium* follows a sinuous path as it swims, ingesting bacteria that it encounters. Whenever it meets unfavorable conditions, like unsuitable acidity or salinity in the water, the *Paramecium* checks its advance, turns and starts in a new direction. Its reaction is purely negative. The *Paramecium* apparently does not seek its food or a favorable environment, but simply avoids unsuitable conditions.

Euglena, also a single-celled organism, exhibits a somewhat greater ability to process information about the environment. *Euglena* has a light-sensitive spot by means of which it can orient itself towards the direction in which the light originates. *Euglena*'s motions are directional; it not only avoids unsuitable environments but it actively seeks suitable ones. An amoeba represents further progress in the same direction; it reacts to light by moving away from it, and also actively pursues food particles.

An increase in the ability to gather and process information about the environment is not a general characteristic of the evolution of life. Progress has occurred in certain evolutionary lines but not in others. Today's bacteria are not more progressive by this criterion than their ancestors of three billion years ago.

The ability to obtain and to process information has progressed little in the plant kingdom. Plants generally react to light and to gravity. The geotropism is positive in the root, but negative in the stem. Plants also grow towards the light; some plants like the sunflower have parts which follow the course of the sun through its daily cycle. Another tropism in plants is the tendency of roots to grow towards water. The response to gravity, to water, and to light is basically due to differential growth rates; a greater elongation of cells takes place on one side of the root or stem than on the other side. Gradients of light, gravity, or moisture are the clues which guide these tropisms. Some plants react also to tactile stimuli.

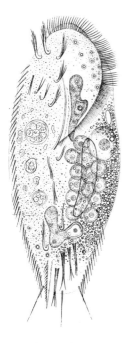

A relative of
Paramecium

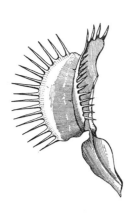

Venus flytrap (reduced; in ready position before snapping shut around a fly) and sundew (enlarged; with sticky tentacles drawn down to digest a mired insect)

Tendrils twine around what they touch; *Mimosa* and carnivorous plants like the Venus flytrap have leaves which close upon being touched.

The ability to obtain and process information about the environment is mediated in multicellular animals by the nervous system. All major groups of animals, except the sponges, have nervous systems. The simplest nervous system occurs in coelenterate hydras, corals, and jellyfishes. Each tentacle of a jellyfish reacts only if it is individually and directly stimulated. There is no coordination of the information gathered by different parts of the animal. Moreover, jellyfishes are unable to learn from experience.

A limited form of coordinated behavior occurs in the echinoderms which comprise the starfishes and sea urchins. Whereas coelenterates possess only an undifferentiated nerve net, echinoderms possess a nerve net, a nerve ring, and radial nerve cords. When the appropriate stimulus is encountered, a starfish reacts with direct and unified actions of the whole body.

The most primitive form of brain occurs in certain organisms like planarian flatworms, which also have numerous sensory cells and eyes without lenses. The information gathered in these sensory cells and organs is processed and coordinated by the central nervous system and the rudimentary brain; a planarian worm is capable of some variability of responses and of some simple learning. That is, the same stimuli will not necessarily always produce the same response.

Planarian flatworms have progressed farther than starfishes in the ability to gather and process information about the environment, and the starfishes have progressed farther than sea anemones and other coelenterates. But none of these organisms has gone very far by this criterion of progress. The most progressive groups of organisms among the invertebrates are the cephalopods and arthropods, but the vertebrates have progressed much farther than any invertebrates.

Among the ancestors of both the arthropods and the vertebrates, there were organisms that, like the sponges, lacked a nervous system. These ancestors evolved through a stage with only a simple network, whereas later stages developed a central nervous system and eventually a rudimentary brain. With further development of the central nervous system and of the brain, the ability to obtain and process information from the outside progressed much farther. The arthropods, which include insects, have complex forms of behavior. Precise visual, chemical, and acoustic signals are obtained and processed by many arthropods, particularly in their search for food and in their selection of mates.

Vertebrates are generally able to obtain and process much more complicated signals and to produce a much greater variety of responses than the arthropods. The vertebrate brain has an enormous number of associative neurons with an extremely complex arrangement. Among the vertebrates, progress in the ability to deal with environmental information is correlated with increase in the size of the cerebral hemispheres and with the appearance and development of the "neopallium." The neopallium is involved in association and coordination of all kinds of impulses from all receptors and brain centers. It first appeared in reptiles. In the mammals it has expanded to become the cerebral cortex, which covers most of the cerebral hemispheres. The cerebral cortex in humans is particularly large, compressed over the hemispheres in a complex pattern of folds and turns. When organisms are measured by their ability to process and obtain information about the environment, mankind is, indeed, the most progressive organism on earth.

There is nothing in the evolutionary process which makes the criterion of progress I have just followed best or more objective than others. It may be useful because it illuminates certain features of the evolution of life. Other criteria may help to discern other features of evolution, and thus be worth examining. Particular organisms will appear more or less progressive depending on the standard that is used to evaluate progress. Mankind is not the most progressive species by many criteria. By some standards, humans are among the bottom rungs of the ladder of life, for example, in the ability to synthesize their own biological materials from inorganic resources.

———

Sessile, unseeing,
the Plant is wholly content
  with the Adjacent.

Mobilised, sighted,
the Beast can tell Here from There
  and Now from Not-Yet.

Talkative, anxious,
Man can picture the Absent
  and Non-Existent.

—*W. H. Auden*

———

Biological Diversity as the Scale of Progress

Edward O. Wilson, Harvard entomologist and biologist extraordinaire, offers another view of biological progress—a view that, in his mind, is not only right but desperately needed to nurture a global environmental ethic. Wilson sees in the broad outline of the history of life factual grounding to support intellectually "what we know in our hearts to be true." Excerpts from The Diversity of Life *(copyright 1992 by Edward O. Wilson) are reprinted by permission of the author and Harvard University Press, Cambridge, Massachusetts.*

Edward O. Wilson   Biological diversity has increased a thousandfold since the early days of
the microbial mats, pulled along by evolutionary progress, measured in
turn by four great steps that mark the passage of eons:

- The origin of life itself, spontaneously from prebiotic organic molecules, about 3.9 to 3.8 billion years ago. The first organisms were single-
  celled and hence microscopic. Stromatolite ecosystems appeared no
  later than 3.5 billion years ago.

- The origin of eukaryotic organisms—"higher organisms"—about 1.8
  billion years ago. Their DNA was enveloped in membranes, and the remainder of the cell contained mitochondria and other well-formed
  organelles. At first eukaryotes were single-celled, in the manner of modern protozoans and the simpler forms of algae, but soon they gave rise
  to more complex organisms composed of many eukaryotic cells organized into tissues and organs.

- The Cambrian explosion, 540 to 500 million years ago. Newly abundant
  macroscopic animals, large enough to be seen with the naked eye,
  evolved in a radiative pattern to create the major adaptive types that
  exist today.

- The origin of the human mind, in later stages of evolution in the genus
  Homo, probably from a million to 100,000 years ago.

Some biologists and philosophers have trouble with that term, "evolutionary progress." The expression is inexact and loaded with humanistic
nuance, granted, but I use it just the same to identify a paradox pivotal
to the understanding of biological diversity. In the strict sense, the concept of progress implies a goal, and evolution has no goal. Goals are not
inherent in DNA. They are not implied by the impersonal forces of natural
selection. Rather, goals are a specialized form of behavior, part of the
outer phenotype that also includes bones, digestive enzymes, and the on-
set of puberty. Once assembled by natural selection, human beings and
other sentient organisms formulate goals as part of their survival strate-
gies. Because goals are the ex-post-facto responses of organisms to the
necessities imposed by the environment, life is ruled by the immediate
past and the present, not by the future. In short, evolution by natural
selection has nothing to do with goals, and so it would seem to have
nothing to do with progress.

And yet there is another meaning of "progress" that does have considerable relevance to evolution. Biological diversity embraces a vast number
of conditions that range from the simple to the complex, with the simple
appearing first in evolution and the more complex later. Many reversals
have occurred along the way, but the overall average across the history of
life has moved from the simple and few to the more complex and numerous. During the past billion years, animals as a whole evolved upward in
body size, feeding and defensive techniques, brain and behavioral com-

Maple Leaves (Shibata Zeshin)

plexity, social organization, and precision of environmental control—in each case farther from the nonliving state than their simpler antecedents did. More precisely, the overall averages of these traits and their upper extremes went up.

An undeniable trend of progressive evolution has been the growth of biodiversity by increasing command of earth's environment. New methods to detect microscopic fossils in billion-year-old sedimentary rocks, chemical analyses of ancient environments, and statistical estimates of the relative abundances of extinct species have allowed geochemists and paleontologists during the past decade to bring this history into sharper focus.

By two billion years before the present, a large fraction of earth's organisms were generating oxygen through photosynthesis. But this element, so vital to life as we know it today, did not accumulate in the water and atmosphere. It was captured by ferrous iron, which dissolves in water and was abundant enough to saturate the early seas. The two elements combined to form ferric oxides, insoluble in water, which settled to the ocean floor. As J. William Schopf neatly summarized the situation, the world rusted.

Denied oxygen by the ferrous sink, the organisms were forced to remain anaerobic. The aerobic pathways of metabolism, which are highly efficient means to obtain and deploy free energy, could at most have evolved as an auxiliary adaptation. By 2.8 billion years ago, the sink had partially filled, and a few local habitats sustained low levels of molecular oxygen. Aerobic organisms, still single-celled prokaryotes, appeared about this time. During the next billion years, oxygen levels rose worldwide to constitute about one percent of the atmosphere. By 1.8 billion years ago, the first eukaryotic organisms appeared: alga-like forms, forerunners of the dominant photosynthesizers of the modern seas. By no later than 600 million years ago, near the end of the Proterozoic era, the first animals evolved. Members of this Ediacaran fauna, named after the Ediacara Hills of South Australia where many of the first specimens were found, were soft-bodied and typically flat. They vaguely resembled jellyfish, annelid worms, and arthropods, and some may have been members of those surviving groups.

Approximately 540 million years ago, near the beginning of the Cambrian period, earliest of the time segments of the Phanerozoic eon in which we now live, a seminal event occurred in the history of life. Animals increased in size and diversified explosively. The supply of free oxygen in the atmosphere was by this time near the 21 percent level of today. The two trends are probably linked, for the simple reason that large, active animals need aerobic respiration and a rich supply of oxygen. Within a few million years, the fossil record held almost every modern phylum of invertebrate animals a millimeter or more in length and possessed of skeletal structures, hence easily preserved and detectable later. A large portion of present-day classes and orders had also come on stage. Thus occurred the Cambrian explosion, the big bang of animal evolution. Bacteria and single-celled organisms had long since attained comparable levels of biochemical sophistication. Now, in a dramatic new radiation, they augmented their niches to include life on the bodies and waste materials of the newly evolved animals. They created a new, microscopic suzerainty of pathogens, symbionts, and decomposers. In broad outline at least, life in the sea attained an essentially modern aspect no later than 500 million years ago.

By this time a strong ozone layer existed as well, screening out lethal short-wave radiation. The intertidal reaches and dry land were safe for

Fossils of Ediacaran soft-bodied organisms. *Upper:* 4.6 cm in length. *Lower:* 2.5 cm in width.

life. By the late Ordovician period, 450 million years ago, the first plants, probably derived from multicellular algae, invaded the land. The terrain was generally flat, lacking mountains, and mild in climate. Animals soon followed: invertebrates of still unknown nature burrowed and tunneled through the primitive soil. Paleontologists have found their trails but still no bodies. Within 50 or 60 million years, early into the Devonian period, the pioneer plants had formed thick mats and low shrubbery widely distributed over the continents. The first spiders, mites, centipedes, and insects swarmed there, small animals truly engineered for life on the land. They were followed by the amphibians, evolved from lobe-finned fishes, and a burst of land vertebrates, relative giants among land animals, to inaugurate the Age of Reptiles. Next came the Age of Mammals and finally the Age of Man, amid continuing tumultuous change at the level of class and order.

By 340 million years before the present, the pioneer vegetation had given way to the coal forests, dominated by towering lycophyte trees, seed ferns, tree horsetails, and a great variety of ferns. Life was close to the attainment of maximal biomass. More organic matter was invested in organisms than ever before. The forests swarmed with insects, including dragonflies, beetles, and cockroaches. By late Paleozoic and early Mesozoic times, close to 240 million years ago, most of the coal vegetation had died out, with the exception of the ferns. Dinosaurs arose among a newly constituted, mostly tropical vegetation, reconstituting the forests and grasslands of the world. The dinosaurs died out during the hegemony of this essentially modern vegetation, at a time when tropical rain forests were assembling the greatest concentration of biodiversity of all time.

For the past 600 million years, the thrust of biodiversity, mass extinction episodes notwithstanding, has been generally upward. The creation of that diversity came slow and hard: 3 billion years of evolution to start the profusion of animals that occupy the seas, another 350 million years to assemble the rain forests in which half or more of the species on earth now live. There was a succession of dynasties. Some species split into two or several daughter species, and their daughters split yet again to create swarms of descendants that deployed as plant feeders, carnivores, free swimmers, gliders, sprinters, and burrowers, in countless motley combinations. These ensembles then gave way by partial or total extinction to newer dynasties, and so on to form a gentle upward swell that carried biodiversity to a peak—just before the arrival of humans. Life had stalled on plateaus along the way, and on five occasions it suffered extinction spasms that took 10 million years to repair. But the thrust was upward. Today the diversity of life is greater than it was 100 million years ago— and far greater than 500 million before that.

Progress, then, is a property of the evolution of life as a whole by almost any conceivable intuitive standard, including the acquisition of

Ten Butterflies (Shibata Zeshin)

goals and intentions in the behavior of animals as well. It makes little sense to judge it irrelevant. Attentive to the adjuration of C. S. Peirce, let us not pretend to deny in our philosophy what we know in our hearts to be true.

The ethical imperative should therefore be, first of all, prudence. We should judge every scrap of biodiversity as priceless while we learn to use it and come to understand what it means to humanity. We should not knowingly allow any species or race to go extinct. And let us go beyond mere salvage to begin the restoration of natural environments, in order to enlarge wild populations and stanch the hemorrhaging of biological wealth. There can be no purpose more enspiriting than to begin the age of restoration, reweaving the wondrous diversity of life that surrounds us.

I turn the handle and the story starts:
Reel after reel is all astronomy,
Till life, enkindled in a niche of sky,
Leaps on the stage to play a million parts.

Life leaves the slime and through all oceans darts;
She conquers earth, and raises wings to fly;
Then spirit blooms, and learns how not to die,
Nesting beyond the grave in others' hearts.

—I turn the handle; other men like me
Have made the film; and now I sit and look
In quiet, privileged like Divinity
To read the roaring world as in a book.
If this thy past, where shall thy future climb,
O Spirit, built of Elements and Time!

—*Julian Huxley*

Julian Huxley's paeans to progress were challenged in his own time by a distinguished paleontologist and evolutionary biologist across the Atlantic. George Gaylord Simpson (1902–1984) conducted most of his research at the American Museum of Natural History in New York City. The arguments and disparate interpretations at the core of the Huxley-Simpson dialectic are at a level that transcends the shifting facts of the field. Paired, their writings become bookends for further inquiry and debate. Moreover, the issue of evolutionary progress is thereby elevated to the ranks of the great—and timeless—questions of philosophy.

Like Julian Huxley, George Gaylord Simpson was a key player during the 1930s and 1940s in the melding of Darwinism with genetics. Paleontologists in Simpson's generation were trained to regard a single "type specimen" of tooth or bone or shell as not just definitive of a species but sufficient for understanding that species. Simpson broke ranks by looking for difference. For Simpson, the more bones the better. A fossil record that displayed variability within species could shed light on the mechanisms of evolution by natural selection. Unless there is variation in a population, after all, there is nothing to select from. This kind of "population" thinking had already transformed the rules of genetics into a substrate suitable for the play of selective forces. Simpson, in similar fashion, made paleontology a partner in the Darwinian paradigm.

Simpson's passion for evolutionary inquiry is revealed in an excerpt from his 1964 book, This View of Life.

George Gaylord Simpson

I do not think that evolution is supremely important because it is my specialty. On the contrary, it is my specialty because it is supremely important. I entered college with the intention of studying literature and becoming a writer, perhaps a poet. I was required to take some laboratory science, and I elected geology, partly because of some previous interest in minerals and partly because Geology 1 was reputed to be a quick and easy way to work off the requirement. Actually, it turned out to be tough because a new professor (who did not last long in that college) demanded an amount of work that most of the students found excessive. But he was an enthusiast, and he imparted to me the thrill of learning things. Here I saw that it was possible to accumulate solid knowledge about the universe, new not only to me but to everybody, and to supply satisfactions that, for me, literary endeavor could not. I switched my major to geology. Slowly I came to feel that although minerals are fascinating, what is really important is life. That made paleontology, the living aspect of geology, my subject in graduate school. Starting then and increasing through the subsequent professional years, a sharpening sense of values showed me that if life is the most important thing about our world, the most important thing about life is its evolution. Thus, by consciously seeking what is most meaningful, I moved from poetry to mineralogy to paleontology to evolution. The transition would have been simpler if I had started with biology, or perhaps even with, say, chemistry; but I think the search would have wound up in the same place.

Simpson, like Julian Huxley, published a number of books and essays that explored the significance of evolutionary biology for the culture at large. Excerpts here are drawn from three of Simpson's works. Passages from The Meaning of Evolution, *copyright 1967/1949 by Yale University Press, are printed with permission of Yale University Press. "The Concept of Progress in Organic Evolution," from the spring 1974 issue of the journal* Social Research *(copyright 1974 Social Research, New York) appears courtesy of the publisher. A fragment is also drawn from "The Nonprevalence of Humanoids," which appeared in 1964 (volume 143) in the journal* Science.

The Idea of Progress

The idea of biological progress is as old as the science of biology and it was already deeply imbedded in pre-evolutionary science. Although its actual historicity and its real relationship to the flow of time were scarcely glimpsed, this concept of progression of life from lower to higher was fundamental both in primitive theology (such as the Semitic Creation myths) and in primitive science (such as that of Aristotle), and it was taken over, more or less as a matter of course, in later pre-evolutionary biology that still stemmed, in the main, from these two sources. Evolution, revealing the development of life as a materially historical process, gave meaning to these older observations and to the almost intuitive concept of progression if not, fully, of progress. The first truly general theory of evolution, that of Lamarck, had as its central feature the very ancient and previously nonevolutionary idea of a sequence of life forms from less to more perfect.

Examination of the actual record of life and of the evolutionary processes as these are now known raises such serious doubts regarding the oversimple and metaphysical concept of a pervasive perfection principle that we must reject it altogether. Yet there is, obviously, progression in the history of life, and if we are to find therein a meaning we are required to consider whether this involves anything that we can agree to call "progress," and if so, its nature and extent.

It is a childish idea—but one deeply ingrained in our thinking, especially on political and social subjects—that change *is* progress. Progression merely in the sense of succession occurs in all things, but one must be hopelessly romantic or unrealistically optimistic to think that its trend is necessarily for the good. We must define progress not merely as movement but as movement in a direction from (in some sense) worse to better, lower to higher, or imperfect to more nearly perfect. A description of what has occurred in the course of evolution will not in itself lead us to the identification of progress unless we decide beforehand that progress must be inherent in these changes. In sober enquiry, we have no real reason to assume, without other standards, that evolution, overall or in any particular case, has been either for better or for worse. Progress can be identified and studied in the history of life only if we first postulate a criterion of progress or can find such a criterion in that history itself.

The criterion natural to human nature is to identify progress as increasing approximation to man and to what man holds good. The criterion is valid and necessary as regards human history, although it carries the still larger obligation of making a defensible and responsible choice among the many and often conflicting things that men have held to be good. Approximation to human status is a reasonable *human* criterion of progress, just as approximation to avian status would be a valid avian criterion or to protozoan status a valid protozoan criterion. It is merely stupid for a man to apologize for being a man or to feel, as with a sense of original sin, that an anthropocentric viewpoint in science or in other fields of thought is automatically wrong. It is, however, even more stupid, and even more common among mankind, to assume that this is the only criterion of progress and that it has a general validity relative to one only among a multitude of possible points of reference.

Evolution and the Fossil Record

Each new organism develops in accordance with a figurative message, coded information, received from its one or two parents. Evolution occurs only if there are changes in that information in the course of generations. Such changes in individuals occur for the most part in two ways: mutations, which introduce new elements into the message, and recombinations, which put these elements into new associations and sequences. The latter sources of variation are sexual, and sexlike processes occur in even the most primitive living organisms.

Since most (but not all) evolutionary changes are adaptive, these processes alone cannot be the whole story. They are necessary for evolution, but something else must also be involved. There must be some interaction between organisms and environment, and from this there must be some kind of feedback into the genetic code. The feedback is by natural selection and it occurs in populations through successive generations, not in individuals in their lifetimes. That is the whole point of natural selection: that it does feed back from environment to genetic code in such a way as to maintain or change the message in adaptive ways. It does this because, by and large, the better adapted organisms have more offspring.

Darwin stated clearly two observations that oppose the idea of progress as an inherent characteristic of evolution. He noted that numerous organisms have persisted for geologically long periods without apparent change. If progress is inherent in evolution, it still did not affect them. Darwin also noted that from any one time in the past it appears that more species (or other taxa) became extinct than survived. For them, a majority, progress was at most a limited thing. Both of those observations have been strongly confirmed and extended since Darwin. There are the so-called living fossils, still extant but little changed from fossil ancestors of tens or hundreds of millions of years ago. Lately, fossils of previously unimaginably great age, from two to three billion (thousand million) years old, have been found. They closely resemble organisms still very much with us (blue-green algae and bacteria), and for these, again, progress seems an inappropriate concept.

The fossil record shows very clearly that there is no central line leading steadily, in a goal-directed way, from a protozoan to man. Instead there has been continual and extremely intricate branching, and whatever course we follow through the branches there are repeated changes both in the rate and in the direction of evolution. Man is the end of one ultimate twig. The housefly, the dog flea, the apple tree, and millions of other kinds of organisms are similarly the ends of others.

Countering Julian Huxley

[Julian] Huxley has cited various objections to the concept of evolutionary progress. He controverted them mainly by admitting and stressing from the start that progress is not general in evolution. Huxley has then gone on with great confidence to propose and define his own concept of evolutionary progress. He based his concept on a "succession of dominant types," naming trilobites (extinct crustaceanlike animals), eurypterids (extinct scorpionlike animals), ostracoderms (fishlike relatives of the lampreys), early true fishes, early insects and amphibians (simultaneously), advanced fishes and advanced insects and reptiles (all three simultaneously), and birds and mammals (simultaneously).

Seeking what all those "dominant types" have in common, Huxley concluded that all their distinguishing characteristics either make for greater control over the environment or make for greater independence from the environment. [In later work] Huxley added various more specific points, including increase in complexity, individuation, capacity for knowledge, emotion, and purpose, and finally a sense of values. Huxley concluded that all previous dominant types are either extinct or in dead

ends and that the possibility for future evolutionary progress exists in man alone.

There can be no greater authority on this subject, and yet Huxley's arguments are open to some grave objections. His basic evidence, the "succession of dominant types," is not really an evolutionary succession. Not in time, because, as he noted, two or more different "types" repeatedly became dominant at the same time. Not in descent, because none of his later "dominant types" was in fact descended from his earlier ones. Not in successively taking over the same sphere, because hardly any two of the "dominant types" really were or are in the same sphere or, broadly, adaptive zone. Birds do not replace fishes, and reptiles do not even replace amphibians.

Each of Huxley's "dominant types" was highly successful for a time, at least, and several of them (advanced insects, advanced fishes, birds, mammals) still are, even though several others are extinct. It is legitimate to inquire what made each so successful, and it is an acceptable concept that the acquisition of characteristics leading to success was progress in each case. However, it is not true that success was due to the same kinds of characteristics in each case or that a real succession can be made of acquisition of increasingly successful adaptations in the cited "dominant types." The point becomes fairly clear if one takes into consideration other groups that seem surely dominant, each in its own sphere, and yet have few or none of the characteristics by which Huxley defined progress. Examples: bacteria, almost certainly the most abundant of all organisms; rodents, surely now a dominant group of land mammals; gastropods (snails and their allies), dominant bottom-livers (benthos) of many waters; not to mention plants such as algae in the waters and flowering plants (angiosperms) on land.

Man Is Not the Measure

Despite his disclaimer, Huxley's concept of progress really is ad hoc and anthropomorphic. To point out that man has more power over nature and also more independence from it merely emphasizes the point: man is a specialist in manipulating his environment (not always to his advantage) and in doing so is more human than monkeys are. That is a legitimate *human* concept of progress. It does not follow and the evidence does not support the proposition that other organisms progressed, became "higher," because of and in proportion to their acquisition of characteristics most highly developed in man.

Our own activities, with the invention of clothing, tools, and the rest, have placed us beyond any other organism in adaptability. The exploitation of that adaptability has occurred, and is still going on, as man adapts further not to one but to a great variety of ways of life. This is progress of a sort and to a degree altogether unique in the history of life.

This development is often expressed as, or I might say "confused with," progressive independence from the environment. It is rather the ability to cope with a greater variety of environments than lessening of dependence on environment as a whole, and the distinction seems real and important. It is also something peculiarly human in its degree and even, in considerable part, in its kind and so must be suspect as a human and not a general or objective criterion of progress in evolution as a whole.

The nonhuman example most often given for progress as increasing independence from the environment is the freeing of the amphibians from dependence on water as a living medium and then of the reptiles from dependence on water in which to lay their eggs. It has been remarkably overlooked that this was only an exchange and not a loss of dependency. The reptile finally became completely dependent on dry land—and water is much more widespread!

Control over environment is still more clearly progress if we agree to consider progress frankly as defined from the human position. At least it leads us to the position that only man is really progressive in the history of life. Actual control of environment, as opposed to the ability merely to move about in search of suitable environments, means of escape from unsuitable ones, or the ability to get along in varied and varying environments, is almost exclusively a human ability. Such things as beaver dams are negligible in comparison and they are isolated marvels in the animal kingdom that will hardly be considered as criteria for evolutionary progress without reference to man. Means of protection from the environment are practically universal among organisms, although they vary greatly in nature. They may exemplify progress, on some other standard, but they do not help particularly to define it.

Progress in Perception

[An] important element in many lines and sorts of evolutionary progress has been "change in the direction of increase in the range and variety of adjustments of the organism to its environment," as [C. J.] Herrick has put it. Note well that this is not at all the same thing as "increased control over and independence of the environment." It is increased awareness and perception of the environment and increased ability to react accordingly.

The lateral line organs of fishes, which are receptors for variations in pressure or movement in a surrounding liquid medium, are lost in terrestrial animals, but the related ear mechanism is adapted for reception of rhythmic pressure variations (sounds) in the air. This mechanism shows marked progress, and in general there is progress, eventually very great, in the central control and reacting mechanisms. Pit vipers, although they live in environments similar to those of other snakes and broadly of many other terrestrial vertebrates, have in the pits for which they are

Head of a fer-de-lance, a poisonous snake found in Central and South America. Between the nostril and the eye is the heat-sensing pit.

named a special sense organ which is a directional receptor for low degrees of heat radiation, a mechanism found in no other animals. The apparatus detects the body heat of small mammals on which these vipers live. This peculiar development certainly counts as progress in perception, but the absence of such an apparatus has no relevance to progress in animals that do not live on warm-blooded prey or that have other adequate means of locating such prey.

This sort of progress, with its innumerable different degrees and many different lines of progression, is more widespread and also more fundamental than such features as independence or control of environment. It is, indeed, progress in perception of and reaction to environment that underlies and makes possible such quite limited degrees of independence and control as have been achieved. It is also evident that this concept of progress is not man-centered but provides a general criterion applicable without necessary reference to the human condition.

No organism has receptors and analyzers for all possible signals from the environment, or for more than a small fraction of these. For the stimuli actually present in his environment, man has about as good a set of receptors as any animal. His associated perceptual, coordinating, and reacting apparatus is incomparably the best ever evolved. Moreover, in the new sort of evolution so characteristic of man, that of social structure and transmitted learning, man has supplemented his organic receptors with external, inorganic receptors of a range and delicacy unknown to any other animal.

Ad Hoc Improvements

Now we come around to what Huxley and indeed all of our authorities have been saying in various ways and within various, sometimes fallible theoretical frameworks. Some organisms *are* better than their ancestors or than some of their relatives at doing certain things in certain ways. Some oysters are better at being oysters than their ancestors. Some trees are better at living on mountain tops than others. We are doubtless better at being men than *Australopithecus* was, although I go along with Haldane far enough to believe that monkeys are better at being monkeys than we would be even if we tried. It is also true that sometimes whole groups have been carried by selection to a point where their great expansion into various adaptive zones became possible, a progressive feature of evolution for which the somewhat confusing term "potential versatility" has lately been introduced. That is the explanation, in unduly broad terms, of the spread of dominant groups from time to time.

With such examples it is perfectly reasonable to say that improvement has factually occurred and that there is therefore evolutionary progress. The progress is, however, ad hoc in every case. Our ancestors' progress was not the oysters', the trees', or the monkeys', nor was theirs ours. Since we are humans, after all, the most interesting and important progress is progress toward us, but let us not mistake this for a general phenomenon.

[There has been, however,] a tendency for life to expand, to fill in all the available spaces in the livable environments, including those created by the process of that expansion itself. This is one possible sort of progress. Accepting it as such, it is the only one that the evidence warrants considering general in the course of evolution. It has been seen that even this, although general, is not invariable. The expansion of life has not been constant and there have even been points where it lost ground temporarily. The general expansion may be considered in terms of the number of individual organisms, of the total bulk of living tissue, or of the gross turnover, metabolism, of substance and energy. It involves all three, and increase in any one is an aspect of progress in this broadest sense.

In summary, evolution is not invariably accompanied by progress, nor does it really seem to be characterized by progress as an essential feature. Progress has occurred within it but is not of its essence. Aside from the broad tendency for the expansion of life, which is also inconstant, there is no sense in which it can be said that evolution *is* progress. Within the framework of the evolutionary history of life there have been not one but many different sorts of progress. Each sort appears not with one single line or even with one central but branching line throughout the course of evolution, but separately in many different lines. These phenomena seem fully consistent with, and indeed readily explained by, the naturalistic theory of evolution. They are certainly inconsistent with the existence of a supernal perfecting principle, with the concept of a goal in evolution, or with control of evolution by autonomous factors, a vital principle common to all forms of life.

Follow the rivers, look for a pass,
or follow the ridges, rise.
There are no eyes on you.
You were kindled from a clot
and washed on the beach like a conch
from one more witless wave.

—Annie Dillard

———

Progression Without Progress

Although Julian Huxley's vision of progress may appeal to the optimists and humanists among us, and while E. O. Wilson's evolutionary outlook carries a special urgency in our era of ecological awakening, biologists today who are willing to speak to the issue most often side with George Gaylord Simpson. John Tyler Bonner, nevertheless, has staked out a middle position. Bonner is a biologist at Princeton whose research has centered on slime molds and other primitive organisms. In his 1989 book, The Evolution of Complexity, *Bonner sieves from the evolutionary record evidence of "progression." But he steers clear of equating progression with progress, and his coarse scale admits no distinction between a perch and a primate. One would be hard pressed, therefore, to find cause for human smugness in Bonner's worldview.*

Bonner's version of evolutionary progression centers on a putative increase in complexity. Previous seers of advance in life history built their cases broadly on the same argument. A hundred years ago, for example, Herbert Spencer portrayed evolution as passage "from the homogeneous to the heterogeneous." Bonner's emphasis on complexity is thus far from original, but it is unique in its quantitative tack. Among other things he develops the theme that the evolutionary increase in complexity, beginning with the earliest forms of microscopic life, is closely correlated with an increase in the size of organisms. Bonner then develops two measurable criteria of the rise of complexity. Excerpts from The Evolution of Complexity *by John Tyler Bonner (copyright 1989 by Princeton University Press) are reprinted here with permission of the author and Princeton University Press.*

Cell Types as a Measure of Complexity

John T. Bonner

Herbert A. Simon (1962) considered complexity, or complex systems, to be made up of a large number of parts that interact, or, as the dictionary says, that are connected together. This immediately makes the point that there are many parts, but I would like to go further than merely to say the parts affect one another. It seems to me that the most interesting thing is that parts are often not only numerous, but frequently they are different in their structure and function. This aspect of complexity gives a division of labor; in fact, this kind of interaction of parts is most characteristic of complex biological systems.

It is also standard to think of complexity as being hierarchical, and Simon makes the important point that such hierarchical structure leads to efficiency of construction. Each part has subparts, and each subpart may be further divided. An obvious example is the mammal, which is made up of organs, which are in turn made up of cells, then molecules, atoms, and finally elementary particles. On the other end of the scale are populations and societies.

If one looks at complexity within organisms, one can equate the subdivision of the innards of the body with a division of labor. There are cells of tissues or organs that specialize in specific functions such as gas exchange, or waste removal, or energy intake, or locomotion, or coordination, and these can be identified by their particular morphologies, which seem to be admirably suited to perform their share of the labor. Probably the easiest way to consider and measure all this division of labor is to count the number of different cell types involved for any particular animal or plant.

Each tissue may seem relatively homogeneous, but it is in fact a composite of numerous cell types. The liver has a number of different kinds of cells, as does the kidney or any other internal organ. Consider the nervous system, including the brain: besides the cells associated with the neurons, such as Schwann cells and glial cells, there are certainly many different kinds of neurons with diverging structural and chemical properties. The problem of counting them with any kind of certainty is extremely difficult—perhaps impossible. We are saved by the fact that for our purposes we need only relative figures, and these can be easily provided for various groups of organisms. A number of authors have tried to do so, and there is reasonable agreement among them.

In the upper end of the complexity scale the taxonomic groups become more inclusive. In the case of vertebrates, it is difficult to imagine that there is any great difference in the number of cell types for fish, amphibians, reptiles, birds, and mammals. What differences there are will be small enough to be drowned by the inaccuracy of our estimates, and therefore they are lumped together. Similar lumping for the same reason has been done for arthropods, molluscs, and annelids. Among plants the difficulty also arises between angiosperms and gymnosperms, and again they have been put together.

A Brief Tour of Cellular Complexity

The smallest organisms with two or three cell types would be spore-forming bacteria. A spore is a truly differentiated cell type and an example of a division of labor. The vegetative cells do the multiplying and growing; the spores resist adverse conditions and will preserve the cell contents for the moment when favorable feeding and growing conditions return. In a fluctuating environment the advantage of such a resistant stage is obvious; this is a major evolutionary step. The smallest spore-forming bacteria weigh between 10^{-11} and 10^{-12} grams.

At the other end of the scale we find two kinds of large prokaryotes that possess two or three cell types. One is to be found among the innumerable examples of blue-green algae *(technically, cyanobacteria)*. Many species, which form long filaments, have periodic cells differing from their neighbors in that their internal cytoplasm appears clear. These so-called heterocysts play a key role in the nitrogen metabolism of the filament and also serve as resistant spores. The other large prokaryote in this category are the myxobacteria. Here the individual cells are true bacteria, but they feed in swarms of ever-increasing size. Ultimately the swarm produces a multicellular fruiting body that consists, in the larger forms, of a nonliving stalk made of congealed slime and, depending on the species, a mass of spores or cysts at the end of the stalk. These fruiting bodies may reach a height of one millimeter and their estimated weight would be roughly 10^{-6} grams.

In the category of four cell types we will include some simple algae and some fungi in which there are male and female gametes, and at least two different kinds of somatic cells. For instance, in the green algae *Volvox*, besides egg and sperm there are the ordinary vegetative cells and certain large cells (gonidia) in the colony that become the asexual colonies of the next generation; they alone are capable of cell division. *Volvox* is a relatively small organism; its weight is roughly 10^{-6} grams.

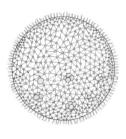

Volvox

At the other end of the scale for algae, consider *Ulva*, the sea lettuce. Besides its gametes, it has two other cell types: the cuboidal cells of the main leaf (thallus) of the plant and the filamentous cells which make up the holdfast *(by which the alga fastens itself to the substrate)*. The size of a mature thallus of *Ulva* may reach that of a human hand, but it is only two cell layers thick, and its estimated weight is approximately five grams.

With approximately seven cell types, we enter a level of complexity where it becomes difficult to give a definitive figure for the number of cell types. If we look to some of the more complex basidiomycetes that form mushrooms, we find that not all the hyphae which make up the thallus are identical; there appears to be a primitive division of labor. Some of the hyphae on the outside are covering cells, and others inside are involved in the support and transport of substances. Also, the structure of the spore-bearing surface (as on a gill in many species of mushroom) is different from that of the rest of the cap and from the stipe that holds the whole structure upright. But the differences are subtle and it is difficult to classify the different hyphae in the various regions into clear-cut cell types. Nevertheless, seven cell types seem to be a reasonable approxi-

Fruiting bodies of various myxobacteria

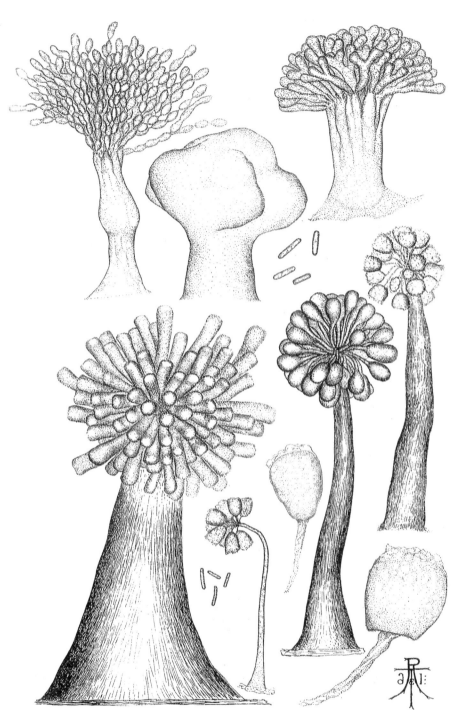

mation. The smallest mushroom will weigh about 10^{-5} grams, and the largest puffball may reach the quite remarkable size of 10^5 grams.

Turning to animals with nine to twelve cell types, we will look specifically at sponges and cnidarians, a group which includes jellyfish. Besides gametes, sponges also have covering cells and collar cells that line their flagellated chambers, and they have a host of different amoeba-like cells that are difficult to distinguish. They contain granules of different colors; one type is responsible for the laying down of the spicules *(glasslike spines that provide support and defense)*. In the case of cnidarians, there is a sharper delineation of cell types such as clear muscle cells, nerve cells, and stinging cells. The distinctions become more difficult when one examines the different cells that line the gut; they are not identical, and some are more directly involved than others in secreting digestive enzymes.

Higher plants, that is, gymnosperms and angiosperms, fall into the category of thirty cell types. In the size range of higher plants, the minimum is in the order of 10^{-4} grams. This range applies, for example, to duckweed (*Lemna*), which floats as small leaves on pond and marsh water, but it has fewer cell types than larger angiosperms. It has very simple roots that protrude below the surface, and a rudimentary stem. Giant sequoias are the largest gymnosperm, and some species of eucalyptus in Australia, which are angiosperms, become almost as large. The difficulty with estimating the weight of such trees, which are about a hundred meters tall, is that, like the giant jellyfish, a large part of their bulk is not alive—in this case, of course, it is dead wood. It is, however, possible to estimate the amount of live cells in a tree, and I am indebted to Dr. John Grace of the University of Edinburgh for showing me how to do this. Approximately ninety percent of the leaves of a tree are made up of live cells, as are nine percent of the trunk, stems, and roots. A rough estimate for a hundred-meter sequoia is 10^7 grams of live cells.

Fifty-five cell types brings us into far more difficult territory—that of higher invertebrates, which include insects and other arthropods, annelid worms, molluscs, and echinoderms *(starfish, sea urchins, and sea cucumbers—all with "spiny skin" and radial symmetry)*. The main point is that it is impossible to make fine distinctions between all these different groups of organisms with respect to their cell-type number. They clearly have more cell types than higher plants. Within this huge spectrum of different kinds of organisms, insects are the smallest. For instance, small wasps called fairy flies, which are a half millimeter long, weigh in the order of magnitude of 10^{-5} grams. The largest complex invertebrate is undoubtedly the giant squid whose weight lies between 10^5 and 10^6 grams. It is the same order of magnitude as an elephant.

Again because fine distinctions are impossible, we shall lump all vertebrates together in the category of more than 120 cell types. No doubt fish have fewer cell types than mammals, but establishing this fact firmly might take a lifetime of sterile investigation. The reason for thinking 120 is an underestimate of the number of cell types is that it is based primarily on morphology, and there is every indication that some cells may be identical morphologically yet quite different in their physiological or biochemical processes. For instance, I am told by an expert in the field that it is possible to recognize ten to twenty morphologically distinct neurons in the central nervous system of a mammal, but that the number could possibly rise to between fifty and one hundred if it included neurons with different neurotransmitter and receptor properties that show no morphological differences.

Therefore, if we were to include nonvisible chemodifferentiation of cells, perhaps vertebrates might have up to five hundred or more cell types. Again, the actual amount is not crucial, but rather the fact that vertebrates have significantly more cell types than invertebrates. The smallest vertebrate is probably some minute species of fish, although there are also exceedingly small frogs; 10^{-2} grams is a reasonable rough value for the minimum size of vertebrates. The largest is clearly the blue whale, which reaches a weight of a hundred tons or 10^8 grams.

Species Diversity as a Measure of Complexity

Let us now consider complexity at the population level. Here one is concerned with communities of organisms in different environments, aquatic or terrestrial. Ecologists often think of a community as a collection of niches, and these niches reflect different ways animals or plants can maintain a place in the community. This is hardly identical to the kind of division of labor within the body in the form of cells and tissues, but it is a close parallel.

If one thinks of niches as slots where organisms can lead a different kind of existence, functioning and behaving in specialized ways, then, if it is not a strict division of labor, it is a division of ways of living. Furthermore, it is often true that the way one type of organism lives is a necessary requirement for the existence of another type, a kind of interdependence clearly analogous to that of the parts within the body of an individual organism. This not only applies in an obvious way to all parasites and obligate symbionts, but in larger ways: for example, all the animals within a community are ultimately dependent on the photosynthetic plants, as the carnivores are dependent on the existence of other animals, and by the same token an epiphyte, such as an orchid, depends on a large tree to hold it up into the air, and so forth.

These ecological comments are meant to make the point that the complexity of a community can be measured in terms of the number of species that exist in it. For a whole community, this is the simplest and most realistic equivalent to the use of cell types as a measure of complexity for an individual organism. The number of species in any one place is called species diversity, or simply diversity. Therefore, in the world within an organism, complexity is measured by the number of cell types, and in the world that surrounds an organism, that is, the world within which it lives, complexity will be measured by the species diversity.

A Brief Tour of Species Diversity

Presumably at some very early stage in earth history the only existing organisms were bacteria. At a later stage there must have been a period in the Precambrian (that is, some time earlier than about 700 million years ago) when a few small invertebrates and some algae coexisted in a community along with protozoa and bacteria. Finally, consider a community today, either on land or in the ocean, where there are organisms ranging in size from bacteria to large trees and mammals, or to huge kelp, large fish, and whales. If one measures complexity in terms of the number of species, clearly these three periods in earth history show a substantial increase in the complexity of biological communities.

In primeval mud, which contained no organisms other than bacteria, the first limiting factor was bound to be the source of energy. Today we know that some prokaryotes use the sun's energy and are photosynthetic, while others get their energy by encouraging a simple chemical reaction such as the conversion of nitrates to nitrites, or sulphur to sulphur dioxide, or iron to ferrous oxide, or even hydrogen into water, and other similar chemical processes. Each of these reactions gives off energy, and these so-called chemosynthetic bacteria grab that energy and use it to build their own proteins, carbohydrates, and other organic cell constituents. Finally, many bacteria can directly use organic substances, such as sugars in the surroundings, but this must have been a relatively late evolutionary development.

Each of these profoundly different ways of obtaining energy is an innovation, a bit of pioneering. By inventing any one of these methods of capturing energy, the bacteria can avoid competition; they make a new world of their own. What we do not know is the historical relationships between the chemosynthetic and the photosynthetic prokaryotes. For our purposes it is only important to think of them as a proliferation of energy-catching techniques. We can assume that initially all prokaryotes used the same method, whatever that ancestral method might have been, and that as competition for the first substrate increased, new energy-trapping inventions were devised. It is more difficult to see how photosynthesis could compete with chemosynthesis, except that again there are two worlds: places where the sun reaches the surface, and places deep in the mud where it cannot.

This is not the place to discuss how the eukaryotic cell arose from earlier prokaryotes. *(All organisms other than bacteria are eukaryotes.)* Whatever the steps, and whatever the forces of selection, it produced diversity in the form of a new type of organism with a radically different kind of cell plan *(in which the genetic material is confined to a nucleus and thereby segregated from the rest of the cell),* one that has been as extraordinarily successful and stable as prokaryote cell construction. One of the important new things about eukaryotic cells is that, besides being able to maintain the ability to perform photosynthesis and take in dissolved foods such as sugars directly, they also invented a new method of capturing energy. It is the engulfing of particulate food; early eukaryotic cells could eat and digest bacteria, thereby becoming predators. The appearance of eukaryotes provided new opportunities for prokaryotes too. Some species of bacteria learned to grow inside eukaryotes and to become symbionts with mutual benefits; in other instances the association is one-sided, that is they became parasites. Each of these steps is pioneering; each means an increase in diversity.

The next major step in evolution might have been the invention of multicellularity. The advantages of multicellularity in predation are especially interesting. Various authors have suggested that the multicellular swarms of bacterial cells in the myxobacteria, and the multinucleate feeding plasmodia of myxomycetes, or true slime molds, are so successful because both secrete massive amounts of extracellular enzymes that dissolve or break down large food particles, making available the simple sugars and amino acids. They are able to do this because they are large and therefore produce high concentrations of enzymes that can break down organic matter which would be unavailable to small, single, feeding cells. Another solution to predation is found in sponges that feed on small particulate food; by creating strong currents they bring in fresh water laden with food and oxygen. Finally, in animals there is the important invention of the gut—first with one opening, as in cnidaria and flatworms, and then with an entrance and exit, as in higher animals.

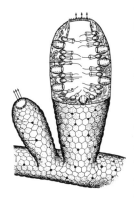

Anatomy of a sponge

An important point about pioneering concerns an organism's entrance into an entirely new kind of habitat, the most conspicuous of which is the conquest of land or of the air. In each case the removal to a new environment was again designed to avoid competition and predation; it was an escape. It has happened many times during the course of evolution. For instance, there are terrestrial bacteria such as the myxobacteria, which make small fruiting bodies that rise up into the air; the same is true of amoebae in the form of slime molds. Terrestrial molds form a very large group, and there are even a number of terrestrial algae. In all of these examples the organism cannot be too far removed from moisture; but in the more complex mosses, ferns, and higher plants, special mechanisms to avoid desiccation have been invented, as well as ways to avoid the need of water for fertilization and sexual reproduction. The same trends are found in animals: many invertebrate groups—such as worms, mollusks, and, most conspicuous of all, insects and other arthropods—have devised wonderful tricks to manage effectively on land. Fi-

nally, the evolution of land-dwelling vertebrates provides an ideal illustration of the point, for again we see ways of preventing water loss and ways of managing fertilization without a water environment to carry the sperm to the egg.

In the conquest of air, insects, birds, and bats are the star performers, although one should not forget the flying reptiles of prehistoric times, or gliding spiders and squirrels. In each instance these separate conquests consisted of a dramatic escape from life on the earth; it permitted greater facility to escape predators, better opportunities to find food, and the opening of a new world. However, we must remember that we are not concerned here with the extraordinary cleverness of the inventions, but rather that they led to the formation of many more new species. Each change in speciation led to further changes, so that diversity bred even greater diversity. We assume that such proliferation of species, as a result of these innovations, must reach some sort of saturation, but who knows what that limit in complexity might be? When all organisms were aquatic, it could have been assumed that saturation existed, but with the opening of land came a whole new wave, and the same is true for the conquest of air. The inventiveness of nature and the diversity that goes with it have been endlessly surprising in the past, and no doubt that tells us something about the future.

And what
if the universe
      is not about us?
  Then what?
     What
       is it about?
   And what
       about *us?*

—May Swenson

"A Thing So Small"

In the 1988 book Evolutionary Progress, *editor Matthew H. Nitecki summarizes the scientific zeitgeist: "The concept of progress has been all but banned from evolutionary biology as being anthropocentric or at best of limited and ambiguous usefulness." That book presents the latest thinking on trends in evolution, case studies of directional movements within one or more of the many stems and twigs of the tree of life. George Gaylord Simpson called these trends "ad hoc improvements," a label that leaves no doubt about their irrelevance to human concerns.*

Stephen Jay Gould's contribution to Nitecki's book is biting, for bat-tling the notion of progress in both popular and scientific thought is a self-appointed duty this Harvard professor undertakes with a vengeance. "Progress," he declaims, "is a noxious, culturally embedded, untestable, nonoperational, intractable idea that must be replaced if we wish to un-derstand the patterns of history." Gould's 1989 best-selling book, Won-derful Life: The Burgess Shale and the Nature of History, *is devoted to this task. Showing that phylum-level diversity in animal forms peaked a half billion years ago at virtually the first appearance of multitissued life, Gould asserts that the only thing determining whether a body plan made it into modernity was luck. Gould further contends that humankind must temper its self-esteem with an awareness of the role that contingency has played in evolution.*

Stephen Jay Gould

If humanity arose just yesterday as a small twig on one branch of a flourishing tree, then life may not, in any genuine sense, exist for us or because of us. Perhaps we are only an afterthought, a kind of cosmic accident, just one bauble on the Christmas tree of evolution.

Gould draws our attention to a poem by Robert Frost titled "Design," which finds in the plight of one tiny moth captured by a spider a mix of contingency and cruelty that makes the poet wonder, "What but design of darkness to appall? / If design govern in a thing so small." Gould then ruminates:

Homo sapiens, I fear, is a "thing so small" in a vast universe, a wildly improbable evolutionary event well within the realm of contingency. Make of such a conclusion what you will. Some find the prospect de-pressing; I have always regarded it as exhilarating, and a source of both freedom and consequent moral responsibility.

The Message of Mass Extinctions

What if Stephen Jay Gould is right? What if the tree of life has been pruned more often by happenstance than by demerit? A close reading of the fossil record of mass extinctions leads David Raup, a paleontologist at the University of Chicago, to ponder:

David Raup

The main question is whether the billions of species that died in the geo-logic past died because they were less fit (bad genes) or merely because they were in the wrong place at the wrong time (bad luck). Do species struggle or gamble for survival? This leads to a question closer to home: Are we here because of a natural superiority (opposable thumbs, big brains, and so on), or are we just plain lucky?

Raup makes a credible case for Lady Luck in his book Extinction: Bad Genes or Bad Luck? *In an introduction to that book, Stephen Jay Gould presents Raup as an unusual kind of paleontologist—one who is "more at home before a computer console than before a dusty drawer of fossils (and he gets his share of flak from traditionalists for this predilection), but he is the acknowledged master of quantitative approaches to the fossil record." Excerpts from* Extinction: Bad Genes or Bad Luck? *(copyright 1991 by David M. Raup) are reprinted here by permission of the author and the publisher, W. W. Norton & Company, Inc.*

All of us grow up with an acquired set of ideas and thoughts about the natural world around us, its history and its future. One idea I think most of us share is that earth is a pretty safe and benevolent place to live—not counting what humans can do to earth and to each other. Earthquakes, hurricanes, and disease epidemics may strike, but on the whole our planet is stable. It is neither too warm nor too cold, the seasons are predictable, and the sun rises and sets on schedule.

Much of our good feeling about planet earth stems from a certainty that life has existed without interruption for three and a half billion years. We have been taught, as well, that most changes in the natural world are slow and gradual. Species evolve in tiny steps over eons; erosion and weathering change our landscape but at an almost immeasurably slow pace. Continents move, as in the present drift of North America away from Europe, but this movement is measured in centimeters per year and will have no practical effect on our lives or those of our children.

Is all this true or merely a fairy tale to comfort us? Is there more to it? I think there is. Almost all species in the past failed. If they died out gradually and quietly and if they deserved to die because of some inferiority, then our good feelings about earth can remain intact. But if they died violently and without having done anything wrong, then our planet may not be such a safe place.

If 99.9 percent of all species that have lived on earth are extinct, it follows that the total of species originations has been virtually the same as the total of species extinctions. Although present biodiversity—the millions of living species—seems high to us, today's biota results from a minor surplus of speciations over extinctions, accumulated over a long time. In view of these figures, it is puzzling that even evolutionary biologists have devoted almost no attention to extinction. Large monographs and textbooks have been written about speciation, and careers have been built around the subject. But extinction has barely been touched. It's a little like a demographer trying to study population growth without considering death rates. Or an accountant interested in credits but not debits.

But interests in science change, and this is happening with extinction. Thanks to a rash proposal by a Nobel physicist, Luis Alvarez, and his colleagues at Berkeley, furious debate has broken out over whether a meteorite impact caused the dinosaur extinction. This has combined with concern over currently endangered species to encourage more people to probe the extinction phenomenon and its role in the history of life.

I should emphasize that extinction is still a very small, cottage industry. It has none of the trappings of big science—nothing comparable to the Supercollider or the Human Genome Project or the Hubble Space Telescope. Yet the questions being asked of extinction are every bit as fundamental and interesting in our ongoing attempt to understand our place in the universe and to answer the ultimate question: Why are we here?

Plausible Causes of Mass Extinctions

When a paleontologist is asked how many mass extinctions there have been, the invariable answer is five: one each in the Ordovician, Devonian, Permian, Triassic, and Cretaceous periods—events known as the Big Five. The best documented of the Big Five came at the end of the Cretaceous. It is often called the K-T—referring to the boundary between the Cretaceous (abbreviated K, to avoid confusion with the Carboniferous and Cambrian periods) and the next younger, or Tertiary (T), period. Because it is the most recent (*65 million years ago*) of the Big Five, its rocks and fossils are the best preserved. Also, sediments from the Cretaceous are widely distributed because it was a time when continents were flooded by shallow seas, leaving a good marine record on the present land surface.

Virtually all plant and animal groups—on land and in the sea—lost species and genera at or near the end of the Cretaceous time. Marine animals suffered the total extinction of 38 percent of their genera; among land animals, the hit was slightly higher. These are big numbers when one considers that in order for a genus to die out, all individuals in all its species must go. Land plants appear to have done a little better, although their fossil record is not good enough for us to be sure.

In the oceans, major losses of species and genera were concentrated among marine reptiles, bony fishes, sponges, snails, clams, ammonites (mollusks distantly related to squids), sea urchins, and foraminifera (single-celled animals usually having a hard skeleton). But no group escaped. Most noteworthy are the large groups—families and orders—that lost all their species. Marine reptiles (plesiosaurs, mosasaurs, and ichthyosaurs), ammonites, and several other once successful groups died out totally. Some had been in decline well before the end of the Cretaceous; others died abruptly. On land, dinosaurs were the most obvious victims, but heavy losses were sustained by a wide variety of other reptiles, mammals, and amphibians. In western North America, fully one-third of all genera of mammals died out at or near the end of the Cretaceous.

Various authors, including news reporters, have made lists of causes that have been suggested for the big mass extinctions. The lists have been intended to show the enormous range of possibilities as well as the silliness of some of them. They do both. "The dinosaurs saw a comet com-

Above: Fossil shells of ammonites (reduced; usually the size of an orange or grapefruit). *Facing page:* Shells of living varieties of microscopic foraminifera.

ing and died of fright." But when the silly ones and the Just So Stories are eliminated, a core of stalwarts remains. I think most of my colleagues would agree on the following as serious candidates, not listed in any particular order:

- climatic change, especially cooling and drying
- sea level rise or fall
- predation
- epidemic disease (a kind of predation)
- competition with other species

(I have purposely left out, though only for the moment, the environmental effects of comet or asteroid impact.)

Each item just listed is reasonable. But I see more than a hint of anthropomorphism here. What are the traditional worries and concerns of people in their own daily lives? The weather, especially cold and lack of rain; water levels (flooding or drying up of rivers and lakes); attacks by wild animals (including insects) or by other people or nations; infectious diseases; and competition (with each other or with other nations). Could it be that the list of probable causes of extinction is simply a list of things that threaten us as individuals? Recall the Four Horsemen of the Apocalypse, representing conquest, war, famine, and plague.

For me to sustain a claim of anthropomorphism, I must be able to draw up a parallel list of plausible but unpopular candidates that do not reflect normal human fears, or at least did not until very recently. Try these:

- chemical poisoning of ocean waters
- changes in atmospheric chemistry
- rocks falling out of the sky
- cosmic radiation
- global volcanic activity
- invasion from outer space

All have been proposed as agents of extinction at one time or another but not taken seriously by the scientific community. The first two are very much on our minds these days, but concern about them is of recent origin and, I submit, has not yet contributed significantly to the canonical list of extinction mechanisms.

The rocks falling out of the sky refers to the idea of a comet/asteroid impact. One reason this notion has caused such a storm of controversy is that it is beyond our experience. Most of us were taught in school that, except for flukes like Meteor Crater, in Arizona, large rocks simply do not fall from the sky. The last three candidates on my list—cosmic radiation, global volcanism, and invasion from space—seem farfetched, and some scientists would try to argue them away on theoretical grounds. But perhaps they merely seem strange (and unlikely) because, like meteorite impacts, they lie outside our experience.

In Praise of Wanton Extinction

Has extinction been a good thing or merely a destructive nuisance barely overridden by the constructive forces of evolution? This is an interesting and tough question without a firm answer.

The important adaptive radiations of the past were evidently made possible by the disappearance of whole groups of species occupying a range of habitats or modes of life. [For example,] many different kinds of organisms have, at one time or another, dominated the framework-building role in tropical reefs. The replacement of one of these kinds by another has generally come only after the elimination of all reefs built by the incumbents. If extinction were a fair game, I think the reef community would have settled down early in Phanerozoic time to be dominated by one kind of frame builder—the best one around at the time. This might or might not have been the optimal organism, but its dominance would have discouraged challenges by new and different organisms. In this scenario, we would still have tropical reefs—and they might have worked perfectly well as ecosystems—but much of evolution's variety would never have been realized.

I conclude, therefore, that extinction is necessary for evolution, as we know it, and that selective extinction that is largely blind to the fitness of the organism (wanton extinction) is most likely to have dominated. As Stephen Jay Gould and others have emphasized, we probably would not be here now if extinction were a completely fair game. NASA and other agencies around the world that search for extraterrestrial life—especially intelligent life—have recognized the importance of extinction in evolution. Twenty years ago, we thought that stable planetary environments would be best for evolution of advanced life. Now NASA is thinking explicitly in terms of planets with enough environmental disturbance to cause extinction and thereby to promote speciation.

Did We Choose a Safe Planet?

The fossil record documents extinctions of many species that were doing fine—until their demise. Are we any different? Is *Homo sapiens* vulnerable to a first strike of natural origin that could do us in despite our proven ability to cope with the normal vicissitudes of nature? Or could a first strike of lesser magnitude—short of total extinction—devastate human civilization as we know it?

Whatever our view of meteorite impact as a cause of extinction in the past, the risk of comet and asteroid impact is part of our present-day environment. But is the risk great enough, compared with other daily hazards, to justify concern? If so, is there anything we can do about it?

Considerable effort has gone into answering these questions. In 1981, the Jet Propulsion Laboratory sponsored a conference at Snowmass, Colorado, chaired by Gene Shoemaker and attended by some of the best solar system astronomers, astrogeologists, engineers, and aeronautics experts. The Snowmass participants concluded that a "civilization-destroying" impact occurs, on average, every 300,000 years. Such an impact would release energy equivalent to more than 8,000 times that of Tunguska and 8 million times that of Hiroshima. (*"Tunguska" refers to a meteor that exploded over Siberia in 1908, leaving no crater, but the shock wave flattened forests for thousands of square miles.*) "Civilization-destroying" means devastation so great that a long period comparable to the Dark Ages (or possibly the Stone Age) would be likely.

If the average spacing of the event is 300,000 years, then the odds are 1:300,000 of civilization's destruction in a given year. For a person living for seventy-five years, the life-time risk is 1:4,000. This gets into the range of other natural and man-made hazards. Clark Chapman and David Morrison, in *Cosmic Catastrophes*, provide some intriguing comparisons. They note that one's chances of experiencing a civilization-destroying impact sometime during a normal lifetime are substantially greater than the chances of dying in an airplane crash, about the same as death through accidental electrocution, and about one-third the risk of being killed in an accidental shooting.

Can there be warning of an impending impact? If so, could anything be done to prevent impact? Would there be time enough? It is estimated that only about 5 percent of large (> 1 km) asteroids with earth-crossing orbits have been discovered. For these, orbits are known well enough that we should have warning times of several years or even decades. But for the 95 percent of earth-crossers that have not been discovered, warning times would be very short, and would depend upon actual sighting of an approaching body.

The normal method of detecting asteroids is to compare photographs of the skyscape taken about forty-five minutes apart. After adjustments for the normal displacement of stars due to the earth's rotation, an asteroid (or comet) is noticeable as a point of light that has shifted between the times when the photographs were taken. Ironically, if the object is coming directly toward earth, no displacement will be seen and the object will probably not be noticed.

It is a catch-22 situation: we will not know how safe (or unsafe) our planet is until more research is done on collision rates—but funding for that research is hard to justify without stronger evidence that the societal risk of impact is serious. We don't know whether we chose a safe planet.

David Raup's essay on the role of mass extinctions in evolution rounds out this section of the book. The perspective he presents is still controversial but has a fair following in the geological community and even in some of the biological sciences now that at least one massive buried crater (on the coast of Mexico's Yucatán Peninsula) has been all but certified as the smoking gun of the great dinosaur extinction. But what about the other four of the Big Five extinctions?

In 1992 Michael Rampino (a geologist at New York University) detected tantalizing clues as to a possible impact site of the biggest extinction of all time—the one whose signature in the fossil record is taken as the end of not only the Permian period but the entire Paleozoic era. Might the unusual structure of the mountains of the Cape Fold Belt in southern Africa be a frozen "tsunami" of a shock wave triggered by a nearby impact? Might that impact have contributed to the breakup of the supercontinent of Gondwanaland, which then connected Africa with South America and Antarctica? And might the thick and extensive deposits of so-called glacial tillite in South America be mislabeled? Might this mishmash of boulders, pebbles, and dust owe not to the waxing and waning of a vast ice sheet but to an impact on a scale too horrible to imagine? And might other records of glacial episodes in times far removed from the Ice Age of our forebears actually signify geologic rather than climatologic holocaust?

Buried craters proposed by Michael Rampino as the site of the end-of-Permian impact. The pair of craters could indicate the splitting of a comet as it entered earth's atmosphere. Possible lines of fracture are drawn, indicating how the impact could have contributed to the breakup of the supercontinent of Gondwanaland.

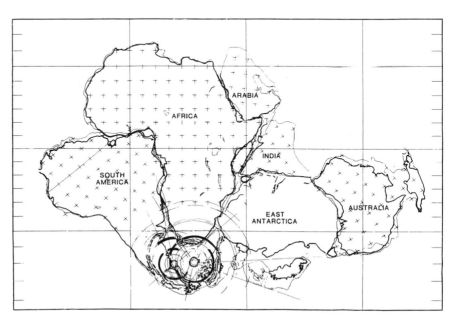

Scientists who suspect or believe that impacts have played a powerful role in evolution are still in the minority. But one thing is certain: this burgeoning research program has the potential to force upon us all a reshaping of worldviews the likes of which Julian Huxley and George Gaylord Simpson could not have imagined.

———

All around me:
      planet, moon, sun, riverbed, marsh:
             grew out of cataclysms galore;
nothing ever sprang whole, stays put.
         I feel the earth beneath my feet
             suddenly shale away;
everywhere I look there's a new disaster,
      and what splintered the mountains
          made gape the pine.

—Diane Ackerman

Foxtail Pine (Paul
Landacre)

II *Tools and Metaphors of Evolution*

Thus, from the war of nature, from famine and death, the most exalted object which we are capable of conceiving, namely, the production of the higher animals, directly follows. There is grandeur in this view of life, with its several powers, having been originally breathed by the Creator into a few forms or into one; and that, whilst this planet has gone cycling on according to the fixed law of gravity, from so simple a beginning endless forms most beautiful and wonderful have been, and are being evolved.

—*Charles Darwin, 1859*

Ecological communities are not simply gladiatorial fields dominated by deadly competition; they are networks of complex interactions, of interdependent self-interests that require mutual adjustment and accommodation with respect to both the other cohabitants and the dynamics of the local ecosystem. The necessity for competition is only one half of a duality, the other half of which includes many opportunities for mutually beneficial co-operation.

—*Peter A. Corning, 1983*

It is adaptations which are due to the animal's behaviour, to its restless exploration of its surroundings, to its initiative in seeking new sources of food when its normal supply fails or becomes scarce through competition, that distinguish the main diverging lines of evolution.

—*Alister Hardy, 1965*

As evident in the first two chapters of this book, the history of life can be read in many ways. Some biologists detect a direction to that story, a meaning, and perhaps even a moral lesson for humankind. Others see in the pattern of speciations and extinctions only a meander, a random walk—our own self-reflective powers and technological prowess an accidental triumph, but not a culmination, and by no means a goal.

Factual knowledge of the history of life may not therefore satisfy the most demanding questers among us. We need to know more. We need to learn what drives that history.

A look at the processes underlying evolution opens up a new level of questioning with its own philosophical gloss. The "tools" of evolution and the metaphors chosen to explain them reveal a diversity of world-views as rich as the cosmological extensions drawn from the history of life.

This three-chapter part begins with the tools and metaphors most familiar to students of biology and to the public at large. The role of strife in crafting evolutionary change is central to the "wedge" metaphor of Charles Darwin and the "arms races" perceived by today's neo-Darwinists.

Darwin's Contribution

George Gaylord Simpson

Darwin was one of history's towering geniuses and ranks with the greatest heroes of man's intellectual progress. He deserves this place first of all because he finally and definitely established evolution as a fact, no longer a speculation or an alternative hypothesis for scientific investigation.

So declares George Gaylord Simpson, who made his own contributions to evolution theory beginning in the 1940s. Naturalists before Charles Darwin had, of course, studied and compared plants and animals discovered on voyages around the world. Many more had pondered the fruits of hundreds of years of intentional breeding of food crops and livestock, and some, like Darwin, had experimented in breeding varieties. (Darwin bred pigeons.) Moreover, in the first half of the nineteenth century when Darwin was developing his ideas, fossil hunting became a professional pursuit; dinosaurs and extinct mammals—shockingly huge—were discovered, especially in the Americas. As naturalist on the voyage of the

Mylodon darwinii, an extinct giant ground sloth

Beagle, *Darwin examined Argentinian fossil beds of extinct mammals, which included bones of the giant ground sloth.*

Charles Darwin was not the first naturalist to look at these clues and to entertain the notion that life had evolved through time. Indeed his grandfather, Erasmus Darwin, had written eloquent (if not scientific) passages proclaiming the fact of evolution. Charles was not even the first to identify a possible mechanism that could drive change in organic forms. But his book, The Origin of Species, *is read to this day for at least three reasons. First, Charles Darwin amassed a stunning volume of evidence to support his case that "descent with modification" had occurred. Second, most biologists today regard the process Darwin identified—"natural selection"—as the primary force acting on genetic recombinations and mutations to shape changes in living forms and behavior. Third, an evolutionary worldview is the bedrock upon which many later scientific theories have been built. Evolution by natural selection is so fundamental to our present understanding that most scientists would find it difficult to explain any aspect of the living world without it.*

Darwin's grand idea has been told and retold by hundreds of scientists. Moreover, the idea itself is still evolving. Writings on the theme of evolution today are full of technical arguments pro and con such major issues as "punctuated equilibrium," "genetic drift," "the neutral theory of molecular evolution." It is astonishing, then, that Darwin's original text remains the clearest and most compelling introduction to the theory of biological evolution.

Darwin published The Origin of Species *in 1859. It encompassed more than two decades of work following his five-year voyage through the Southern Hemisphere. That voyage brought Darwin not only grand insights but a devastating (and undiagnosed) illness that plagued him for the rest of his life. How little was known then about the dangers of tropical diseases can be seen in this passage from Darwin's journal: "Numbers of fireflies were hovering about, and the mosquitoes were very troublesome. I exposed my hand for five minutes, and it was soon black with them; I do not suppose there could have been less than fifty, all busy sucking."*

Many years later, reflecting on the habits that brought him success, Darwin concludes, "Of these, the most important have been—the love of science—unbounded patience and long reflecting over any subject—industry in observing and collecting facts—and a fair share of invention as well as common sense."

Darwin's Metaphorical Style

As a writer, Darwin was a master of metaphor. How else does one convey a novel idea, except in the context of the familiar? Aristotle observed,

"A good metaphor implies an intuitive perception of the similarity in dissimilars."

Clearly, Darwin derived many of his metaphors from the ethos of laissez-faire economics, as the industrial revolution was then sweeping through his native England. In later editions of The Origin of Species *Darwin calls natural selection "survival of the fittest"—a phrase borrowed from philosopher and sociologist Herbert Spencer. Another potent metaphor is the wedge. Darwin portrays the biological world as "a yielding surface, with ten thousand sharp wedges packed close together and driven inwards by incessant blows, sometimes one wedge being struck and then another with greater force." For every wedge that is driven deeper, another, necessarily, will be edged out. Competition thus plays a pivotal role in the evolution of life.*

The Origin of Species *is unquestionably a great work, but is it a good read? Jacques Barzun, a critic of scientific materialism, attacks Darwin's style as well as his message. In a 1941 book,* Darwin, Marx, Wagner: Critique of a Heritage, *Barzun assails Darwin's "endless shufflings with words." He continues:*

Jacques Barzun

Many handbooks of writing quote excerpts from Darwin as models of scientific prose. They are samples rather than models, for they could be matched by equally good ones in hundreds of other writers. They do not in any case justify the extravagant opinion occasionally heard that the *Origin of Species* is a masterpiece of English literature—an opinion which can come only from revelation by faith and not from the experience of reading the work.

Stanley E. Hyman, literary critic, disagrees. In his 1974 book, The Tangled Bank: Darwin, Marx, Frazer and Freud as Imaginative Writers, *he writes:*

Stanley E. Hyman

Darwin's frankness about objections and unsolved problems gives the book the character of a mythic quest. It is remarkable enough, I suppose, to find that *The Origin of Species* is a work of literature, with the structure of tragic drama and the texture of poetry. Probably, however, the work is even more ambitious than that, and constitutes something like a sacred writing, a scripture. Some of the book's prophetic quality comes from Darwin's predominant tone of personal testimony: I was there, I saw it, this happened to me.

I admit to leaning more in the direction of Hyman than Barzun. I remember distinctly four days of the summer of 1988. Reading Darwin for the first time, outdoors, with fuchsia ablaze and only the occasional buzz of a hummingbird to break my concentration, I was in intellectual heaven. Having to choose just a few morsels for presentation here has been a sadness as well as a challenge. But judge the literary and intellectual merits of Darwin's words for yourself. Bear in mind that all head-

ings have been added and some of the text appears out of order. A concordance at the end of the book will allow you to match any of these fragments to pages in a widely available, sixth edition of Darwin's classic text.

The Origin of Species

Charles Darwin

When on board the H.M.S. 'Beagle', as naturalist, I was much struck with certain facts in the distribution of the organic beings inhabiting South America, and in the geological relations of the present to the past inhabitants of that continent. These facts seemed to throw some light on the origin of species—that mystery of mysteries, as it has been called by one of our greatest philosophers. On my return home, it occurred to me, in 1837, that something might perhaps be made out on this question by patiently accumulating and reflecting on all sorts of facts which could possibly have any bearing on it. After five years' work I allowed myself to speculate on the subject, and drew up some short notes; these I enlarged in 1844 into a sketch of the conclusions, which then seemed to me probable: from that period to the present day I have steadily pursued the same object. I hope that I may be excused for entering on these personal details, as I give them to show that I have not been hasty in coming to a decision.

My work is now (1859) nearly finished; but as it will take me many more years to complete it, and as my health is far from strong, I have been urged to publish this Abstract. I have more especially been induced to do this, as Mr. Wallace, who is now studying the natural history of the Malay archipelago, has arrived at almost exactly the same general conclusions that I have on the origin of species. *(Alfred Russel Wallace is credited as co-founder of the theory of evolution by natural selection. Indeed, it was because Wallace sent Darwin a draft of a short scientific paper with an uncanny resemblance to Darwin's own long-simmering ideas that Darwin finally consented to publish his work-in-progress. Darwin's "abstract," by the way,* The Origin of Species, *is longer than the book you now hold.)*

In considering the Origin of Species, it is quite conceivable that a naturalist, reflecting on the mutual affinities of organic beings, on their embryological relations, their geographical distribution, geological succession, and other such facts, might come to the conclusion that species had not been independently created, but had descended, like varieties, from other species. Nevertheless, such a conclusion, even if well founded, would be unsatisfactory, until it could be shown how the innumerable species inhabiting this world have been modified, so as to acquire that perfection of structure and coadaptation which justly excites our admiration.

How have all those exquisite adaptations of one part of the organisation to another part, and to the conditions of life, and of one organic being to another being, been perfected? We see these beautiful co-adaptations most plainly in the woodpecker and the mistletoe; and only a little less plainly in the humblest parasite which clings to the hairs of a quadruped or feathers of a bird; in the structure of the beetle which dives through the water; in the plumed seed which is wafted by the gentlest breeze; in short, we see beautiful adaptations everywhere and in every part of the organic world.

It is therefore of the highest importance to gain a clear insight into the means of modification and co-adaptation.

Geometrical Rate of Increase

Linnaeus has calculated that if an annual plant *(an annual plant dies at the end of a single growing season)* produced only two seeds—and there is no plant so unproductive as this—and their seedlings next year produced two, and so on, then in twenty years there should be a million plants. There is no exception to the rule that every organic being naturally increases at so high a rate, that, if not destroyed, the earth would soon be covered by the progeny of a single pair. Even slow-breeding man has doubled in twenty-five years, and at this rate, in less than a thousand years, there would literally not be standing-room for his progeny.

But we have better evidence on this subject than mere theoretical calculations, namely, the numerous recorded cases of the astonishingly rapid increase of various animals in a state of nature, when circumstances have been favourable to them during two or three following seasons. Still more striking is the evidence from our domestic animals of many kinds which have run wild in several parts of the world; if the statements of the rate of increase of slow-breeding cattle and horses in South America, and latterly in Australia, had not been well authenticated, they would have been incredible. So it is with plants; cases could be given of introduced plants which have become common throughout whole islands in a period of less than ten years. Several of the plants, such as the cardoon and a tall thistle, which are now the commonest over the plains of La Plata *(in Argentina)* clothing square leagues of surface almost to the exclusion of every other plant, have been introduced from Europe; and there are plants which now range in India, as I hear from Dr. Falconer, from Cape Comorin to the Himalaya, which have been imported from America since its discovery.

In such cases, and endless others could be given, no one supposes that the fertility of the animals or plants has been suddenly and temporarily increased in any sensible degree. The obvious explanation is that the conditions of life have been highly favourable, and that there has consequently been less destruction of old and young, and that nearly all the

young have been enabled to breed. Their geometrical ratio of increase, the result of which never fails to be surprising, simply explains their extraordinarily rapid increase and wide diffusion in their new homes.

The only difference between organisms which annually produce eggs or seeds by the thousand and those which produce extremely few is that the slow-breeders would require a few more years to people, under favourable conditions, a whole district, let it be ever so large. The Fulmar petrel lays but one egg, yet it is believed to be the most numerous bird in the world. One fly deposits hundreds of eggs, and another, like the hippobosca, a single one; but this difference does not determine how many individuals of the two species can be supported in a district.

A large number of eggs is of some importance to those species which depend on a fluctuating amount of food, for it allows them rapidly to increase in number. But the real importance of a large number of eggs or seeds is to make up for the much destruction at some period of life; and this period in the great majority of cases is an early one. If an animal can in any way protect its own eggs or young, a small number may be produced, and yet the average stock be fully kept up; but if many eggs or young are destroyed, many must be produced, or the species will become extinct. It would suffice to keep up the full number of a tree, which lived on an average for a thousand years, if a single seed were produced once in a thousand years, supposing that this seed were never destroyed, and could be ensured to germinate in a fitting place. So that, in all cases, the average number of any animal or plant depends only indirectly on the number of its eggs or seeds.

The Struggle for Existence

Every being, which during its natural lifetime produces several eggs or seeds, must suffer destruction during some period of its life, and during some season or occasional year, otherwise, on the principle of geometrical increase, its numbers would quickly become so inordinately great that no country could support the product. Hence, as more individuals are produced than can possibly survive, there must in every case be a struggle for existence, either one individual with another of the same species, or with the individuals of distinct species, or with the physical conditions of life. It is the doctrine of Malthus applied with manifold force to the whole animal and vegetable kingdoms; for in this case there can be no artificial increase of food, and no prudential restraint from marriage. *(Thomas Malthus had produced in 1798 a widely read book,* An Essay on the Principle of Population, *in which he pointed out that human population tends to increase faster than its food supply, thus prompting a "struggle for existence." Historians believe that Malthus had a profound influence on Darwin's ideas.)*

The term, Struggle for Existence: I should premise that I use this term in a large and metaphorical sense, including dependence of one being on another, and including (which is more important) not only the life of the individual, but success in leaving progeny. Two canine animals, in a time of dearth, may be truly said to struggle with each other which shall get food and live. But a plant on the edge of a desert is said to struggle for life against the drought, though more properly it should be said to be dependent on the moisture. A plant which annually produces a thousand seeds, of which only one of an average comes to maturity, may be more truly said to struggle with the plants of the same and other kinds which already clothe the ground. The mistletoe is dependent on the apple and a few other trees, but can only in a far-fetched sense be said to struggle with these trees, for, if too many of these parasites grow on the same tree, it languishes and dies. But several seedling mistletoes, growing close together on the same branch, may more truly be said to struggle with each other. As the mistletoe is disseminated by birds, its existence depends on them; and it may methodically be said to struggle with other fruit-bearing plants, in tempting the birds to devour and thus disseminate its seeds. In these several senses, which pass into each other, I use for convenience' sake the general term of Struggle for Existence.

The amount of food for each species of course gives the extreme limit to which each can increase; but very frequently it is not the obtaining food, but the serving as prey to other animals, which determines the average numbers of a species. Thus, there seems to be little doubt that the stock of partridges, grouse, and hares on any large estate depends chiefly on the destruction of vermin. If not one head of game were shot during the next twenty years in England, and, at the same time, if no vermin were destroyed, there would, in all probability, be less game than at present, although hundreds of thousands of game animals are now annually shot. On the other hand, in some cases, as with the elephant, none are destroyed by beasts of prey; for even the tiger in India most rarely dares to attack a young elephant protected by its dam.

Climate plays an important part in determining the average number of a species, and periodical seasons of extreme cold or drought seem to be the most effective of all checks. I estimated (chiefly from the greatly reduced numbers of nests in the spring) that the winter of 1854–5 destroyed four-fifths of the birds in my own grounds; and this is a tremendous destruction, when we remember that ten percent is an extraordinarily severe mortality from epidemics with man. The action of climate seems at first sight to be quite independent of the struggle for existence; but in so far as climate chiefly acts in reducing food, it brings on the most severe struggle between the individuals, whether of the same or of distinct species, which subsist on the same kind of food. Even when climate, for instance extreme cold, acts directly, it will be the least vigorous individu-

Dryptosaurus (Charles Knight)

als, or those which have got least food through the advancing winter, which will suffer most.

Eggs or very young animals seem generally to suffer most, but this is not invariably the case. With plants there is a vast destruction of seeds, but, from some observations which I have made it appears that the seedlings suffer most from germinating in ground already thickly stocked with other plants. Seedlings, also, are destroyed in vast numbers by various enemies; for instance, on a piece of ground three feet long and two wide, dug and cleared, and where there could be no choking from other plants, I marked all the seedlings of our native weeds as they came up, and out of 357 no less than 295 were destroyed, chiefly by slugs and insects. If turf which has long been mown be let to grow, the more vigorous plants gradually kill the less vigorous, though fully grown plants; thus out of twenty species growing on a little plot of mown turf (three feet by four) nine species perished, from the other species being allowed to grow up freely.

The dependency of one organic being on another, as of a parasite on its prey, lies generally between beings remote in the scale of nature. This is likewise sometimes the case with those which may be strictly said to struggle with each other for existence, as in the case of locusts and grass-feeding quadrupeds. But the struggle will almost invariably be most severe between individuals of the same species, for they frequent the same districts, require the same food, and are exposed to the same dangers. In the case of varieties of the same species, the struggle will generally be almost equally severe, and we sometimes see the contest soon decided: for instance, if several varieties of wheat be sown together, and the mixed seed be resown, some of the varieties which best suit the soil or climate, or are naturally the most fertile, will beat the others and so yield more seed, and will consequently in a few years supplant the other varieties. To keep up a mixed stock of even such extremely close varieties as the variously-coloured sweet peas, they must be each year harvested separately, and the seed then mixed in due proportion, otherwise the weaker kinds will steadily decrease in number and disappear.

Nothing is easier to admit in words the truth of the universal struggle for life, or more difficult—at least I have found it so—than constantly to bear this conclusion in mind. Yet unless it be thoroughly engrained in the mind, the whole economy of nature, with every fact on distribution, rarity, abundance, extinction, and variation, will be dimly seen or quite misunderstood. We behold the face of nature bright with gladness, we often see superabundance of food; we do not see or we forget, that the birds which are idly singing round us mostly live on insects or seeds, and are thus constantly destroying life; or we forget how largely these songsters, or their eggs, or their nestlings, are destroyed by birds and beasts of prey; we do not always bear in mind, that, though food may be now superabundant, it is not so at all seasons of each recurring year.

Complex Relations

Many cases are on record showing how complex and unexpected are the checks and relations between organic beings, which have to struggle together in the same country. I will give only a single instance, which, though a simple one, interested me. In Staffordshire, on the estate of a relation, where I had ample means of investigation, there was a large and extremely barren heath, which had never been touched by the hand of man; but several hundred acres of exactly the same nature had been enclosed twenty-five years previously and planted with Scotch fir. The change in the native vegetation of the planted part of the heath was most remarkable, more than is generally seen in passing from one quite different soil to another: not only the proportional numbers of the heath-plants were wholly changed, but twelve species of plants (not counting grasses and carices) flourished in the plantations, which could not be found on the heath. The effect on the insects must have been still greater, for six insectivorous birds were very common in the plantations, which were not to be seen on the heath; and the heath was frequented by two or three distinct insectivorous birds. Here we see how potent has been the effect of the introduction of a single tree, nothing whatever else having been done, with the exception of the land having been enclosed, so that cattle could not enter.

But how important an element enclosure is, I plainly saw near Farnham, in Surrey. Here there are extensive heaths, with a few clumps of old Scotch firs on the distant hilltops: within the last ten years large spaces have been enclosed, and self-sown firs are now springing up in multitudes, so close together that all cannot live. When I ascertained that these young trees had not been sown or planted, I was so much surprised at their numbers that I went to several points of view, whence I could examine hundreds of acres of the unenclosed heath, and literally I could not see a single Scotch fir, except the old planted clumps. But on looking closely between the stems of the heath, I found a multitude of seedlings and little trees which had been perpetually browsed down by the cattle. In one square yard, at a point some hundred yards distant from one of the old clumps, I counted thirty-two little trees; and one of them, with twenty-six rings of growth, had, during many years tried to raise its head above the stems of the heath, and had failed. No wonder that, as soon as the land was enclosed, it became thickly clothed with vigorously growing young firs. Yet the heath was so extremely barren and so extensive that no one would ever have imagined that cattle would have so closely and effectually searched it for food.

Here we see that cattle absolutely determine the existence of the Scotch fir; but in several parts of the world insects determine the existence of cattle. Perhaps Paraguay offers the most curious instance of this; for here neither cattle nor horses nor dogs have ever run wild, though they swarm southward and northward in a feral state; and Azara and Rengger have shown that this is caused by the greater number in Paraguay of a certain fly, which lays its eggs in the navels of these animals when first born. The increase of these flies, numerous as they are, must be habitually checked by some means, probably by other parasitic insects. Hence, if certain insectivorous birds were to decrease in Paraguay, the parasitic insects would probably increase; and this would lessen the number of the navel-frequenting flies—then cattle and horses would become feral, and this would certainly greatly alter (as indeed I observed in parts of South America) the vegetation: this again would largely affect the insects; and this, as we have just seen in Staffordshire, the insectivorous birds, and so onwards in ever-increasing circles of complexity.

Not that under nature the relations will ever be as simple as this. Battle within battle must be continually recurring with varying success; and yet in the long-run the forces are so nicely balanced, that the face of nature remains for long periods of time uniform, though assuredly the merest trifle would give the victory to one organic being over another. Nevertheless, so profound is our ignorance, and so high our presumption, that we marvel when we hear of the extinction of an organic being; and as we do not see the cause, we invoke cataclysms to desolate the world, or invent laws on the duration of the forms of life!

When we look at the plants and bushes clothing an entangled bank, we are tempted to attribute their proportional numbers and kinds to what we call chance. But how false a view is this! Every one has heard that when an American forest is cut down a very different vegetation springs up; but it has been observed that ancient Indian ruins in the Southern United States, which must formerly have been cleared of trees, now display the same beautiful diversity and proportion of kinds as in the surrounding virgin forest. What a struggle must have gone on during long centuries between the several kinds of trees each annually scattering its seeds by the thousand; what war between insect and insect—between insects, snails, and other animals with birds and beasts of prey—all striving to increase, all feeding on each other, or on the trees, their seeds and seedlings, or on the other plants which first clothed the ground and thus checked the growth of the trees! Throw up a handful of feathers, and all fall to the ground according to definite laws; but how simple is the problem where each shall fall compared to that of the action and reaction of the innumerable plants and animals which have determined, in the course of centuries, the proportional numbers and kinds of trees now growing on the old Indian ruins!

A corollary of the highest importance may be deduced from the foregoing remarks, namely, that the structure of every organic being is related, in the most essential yet often hidden manner, to that of all the other organic beings with which it comes into competition for food or residence, or from which it has to escape, or on which it preys. This is obvious in the structure of the teeth and talons of the tiger; and in that of the legs and claws of the parasite which clings to the hair on the tiger's body. But in the beautifully plumed seed of the dandelion, and in the flattened and fringed legs of the water-beetle, the relation seems at first confined to the elements of air and water. Yet the advantage of plumed seeds no doubt stands in the closest relation to the land being already thickly clothed with other plants; so that the seeds may be widely distributed and fall on unoccupied ground. In the water-beetle, the structure of its legs, so well adapted for diving, allows it to compete with other aquatic insects, to hunt for its own prey, and to escape serving as prey to other animals.

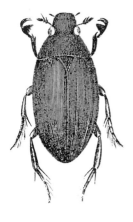

Water beetle

It is good thus to try in imagination to give to any one species an advantage over another. Probably in no single instance should we know what to do. This ought to convince us of our ignorance on the mutual relations of all organic beings; a conviction as necessary as it is difficult to acquire. All that we can do is to keep steadily in mind that each organic being is striving to increase in a geometrical ratio; that each at some period of its life, during some season of the year, during each generation or at intervals, has to struggle for life and to suffer great destruction. When we reflect on this struggle, we may console ourselves with the full belief, that the war of nature is not incessant, that no fear is felt, that death is generally prompt, and that the vigorous, the healthy, and the happy survive and multiply.

Natural Selection

In looking at Nature, it is most necessary to keep the foregoing considerations always in mind—never to forget that every single organic being may be said to be striving to the utmost to increase its numbers; that each lives by a struggle at some period of its life; that heavy destruction inevitably falls either on the young or old, during each generation or at recurrent intervals. Lighten any check, mitigate the destruction ever so little, and the number of the species will almost instantaneously increase to any amount.

Owing to this struggle, variations, however slight and from whatever cause proceeding, if they be in any degree profitable to the individuals of a species, in their infinitely complex relations to other organic beings and to their physical conditions of life, will tend to the preservation of such individuals, and will generally be inherited by the offspring. The offspring, also, will thus have a better chance of surviving, for, of the many individuals of any species which are periodically born, but a small number can survive. I have called this principle, by which each slight variation, if useful, is preserved, by the term Natural Selection, in order to mark its relation to man's power of selection. But the expression often used by Mr. Herbert Spencer of the Survival of the Fittest is more accurate, and is sometimes equally convenient.

Several writers have misapprehended or objected to the term Natural Selection. *(Remember: these excerpts are drawn from the sixth edition of* The Origin of Species, *published a dozen years after the original.)* Some have even imagined that natural selection induces variability, whereas it implies only the preservation of such variations as arise and are beneficial to the being under its conditions of life. No one objects to agriculturalists speaking of the potent effects of man's selection; and in this case, the individual differences given by nature, which man for some object selects, must of necessity first occur. Others have objected that the term selection implies conscious choice in the animals which become modified; and it has even been urged that, as plants have no volition, natural selection is not applicable to them! In the literal sense of the word, no doubt, natural selection is a false term; but who ever objected to chemists speaking of the elective affinities of the various elements?—and yet an acid cannot strictly be said to elect the base with which it in preference combines.

It has been said that I speak of natural selection as an active power or Deity; but who objects to an author speaking of the attraction of gravity as ruling the movements of the planets? Every one knows what is meant

and is implied by such metaphorical expressions; and they are almost necessary for brevity. So again it is difficult to avoid personifying the word Nature; but I mean by Nature, only the aggregate action and product of many natural laws, and by laws the sequence of events as ascertained by us. With a little familiarity such superficial objections will be forgotten.

We have seen that man by selection can certainly produce great results, and can adapt organic beings to his own uses, through the accumulation of slight but useful variations, given to him by the hand of Nature. But Natural Selection is a power incessantly ready for action, and is as immeasureably superior to man's feeble efforts, as the works of Nature are to those of Art. Slow though the process of selection may be, if feeble man can do much by artificial selection, I can see no limit to the amount of change, to the beauty and complexity of the coadaptations between all organic beings, one with another and with their physical conditions of life, which may have been effected in the long course of time through nature's power of selection, that is by the survival of the fittest.

It is interesting to contemplate a tangled bank, clothed with many plants of many kinds, with birds singing on the bushes, with various insects flitting about, and with worms crawling through the damp earth, and to reflect that these elaborately constructed forms, so different from each other, and dependent upon each other in so complex a manner, have all been produced by laws acting around us. These laws, taken in the largest sense, being Growth and Reproduction; Inheritance which is almost implied by reproduction; Variability from the indirect action of the conditions of life, and from use and disuse *("use and disuse" is now a discredited mechanism for inheritance)*; a Ratio of Increase so high as to lead to a Struggle for Life, and as a consequence to Natural Selection, entailing Divergence of Character and the Extinction of less-improved forms.

Thus, from the war of nature, from famine and death, the most exalted object which we are capable of conceiving, namely, the production of the higher animals, directly follows. There is grandeur in this view of life, with its several powers, having been originally breathed by the Creator into a few forms or into one; and that, whilst this planet has gone cycling on according to the fixed law of gravity, from so simple a beginning endless forms most beautiful and wonderful have been, and are being, evolved. *(In the first edition, the phrase "by the Creator" did not appear.)*

———

I am a frayed and nibbled survivor
In a fallen world, and I am getting along.
I am aging and eaten and have done my share
Of eating too. I am not washed and beautiful,

In control of a shining world in which everything fits,
But instead am wandering awed about on a splintered wreck
I've come to care for, whose gnawed trees breathe
A delicate air, whose bloodied and scarred creatures
Are my dearest companions, and whose beauty beats and shines
Not *in* its imperfections but overwhelmingly in spite of them.

—Annie Dillard

———

Evolutionary Arms Races

Charles Darwin used the example of wedges to illumine his theory of natural selection. He also regarded Herbert Spencer's coinage "survival of the fittest" an apt metaphor for the interactions that drive "descent with modification"—which, today, we call evolution. In the past several decades, evolutionary biologists have invented a new metaphor, "evolutionary arms races," for the dynamic that arises between species that are, in some sense, mutual enemies.

Richard Dawkins is a major contributor to research on evolutionary arms races. An Oxford University biologist trained in ethology, the evolutionary aspects of animal behavior, Dawkins is best known for his "selfish gene theory" and his 1976 book, The Selfish Gene. *(See chapter 8.) Excerpts in this chapter are drawn from his third book. Passages from* The Blind Watchmaker *by Richard Dawkins (copyright 1986 by Richard Dawkins) are reprinted here by permission of the author and the publishers: W. W. Norton & Company, Inc. and Sterling Lord Literistic, Inc.*

Richard Dawkins   The human mind is an inveterate analogizer. We are compulsively drawn to see meaning in slight similarities between very different processes. I spent much of a day in Panama watching two teeming colonies of leaf-cutter ants fighting, and my mind irresistibly compared the limb-strewn battlefield to pictures I had seen of Passchendaele. I could almost hear the guns and smell the smoke. Shortly after my first book, *The Selfish Gene*, was published, I was independently approached by two clergymen, who both had arrived at the same analogy between ideas in the book and the doctrine of original sin. Darwin applied the idea of evolution in a discriminating way to living organisms changing in body form over countless generations. His successors have been tempted to see evolution in everything; in the changing form of the universe, in developmental 'stages' of human civilizations, in fashions in skirt lengths.

Sometimes such analogies can be immensely fruitful, but it is easy to push analogies too far, and get overexcited by analogies that are so tenuous as to be unhelpful or even downright harmful. I have become accustomed to receiving my share of crank mail, and have learned that one of the hallmarks of futile crankiness is overenthusiastic analogizing. On the other hand, some of the greatest advances in science have come about because some clever person spotted an analogy between a subject that was already understood, and another still mysterious subject. The trick is to strike a balance between too much indiscriminate analogizing on the one hand, and a sterile blindness to fruitful analogies on the other. The successful scientist and the raving crank are separated by the quality of their inspirations. But I suspect that this amounts, in practice, to a difference, not so much in ability to notice analogies as in ability to reject foolish analogies and pursue helpful ones.

In the world of nations, when two enemies each progressively improve their weaponry in response to the other side's improvements, we speak of an 'arms race'. The evolutionary analogy is close enough to justify borrowing the term, and I make no apology to my pompous colleagues who would purge our language of such illuminating images. I regard arms races as of the utmost importance because it is largely arms races that have injected such 'progressiveness' as there is in evolution. For, contrary to earlier prejudices, there is nothing inherently progressive about evolution.

Arms races consist of the improvement in one lineage's (say prey animals') equipment to survive, as a direct consequence of improvement in another (say predators') lineage's evolving equipment. There are arms races wherever individuals have enemies with their own capacity for evolutionary improvement. We can use the general term 'enemies' of a species, to mean other living things that work to make life difficult. Lions are enemies of zebras. It may seem a little callous to reverse the statement to 'Zebras are enemies of lions'. The role of the zebra in the relationship seems too innocent and wronged to warrant the pejorative 'enemy'. But individual zebras do everything in their power to resist being eaten by lions, and from the lions' point of view this is making life harder for them. If zebras and other grazers all succeeded in their aim, the lions would die of starvation. So by our definition zebras are enemies of lions. Parasites such as tapeworms are enemies of their hosts, and hosts are enemies of parasites since they tend to evolve measures to resist them. Herbivores are enemies of plants, and plants are enemies of herbivores, to the extent that they manufacture thorns, and poisonous or nasty-tasting chemicals.

Lineages of animals and plants will, in evolutionary time, 'track' changes in their enemies no less assiduously than they track changes in average weather conditions. Evolutionary improvements in cheetah weaponry and tactics are, from the gazelles' point of view, like a steady worsening of the climate, and they are tracked in the same kind of way. But there is one enormously important difference between the two. The weather changes over the centuries, but it does not change in a specifically malevolent way. It is not out to 'get' gazelles.

The tendency for carnivores to get progressively 'better' would soon run out of steam were it not for the parallel tendency in the prey. One complication is that a given species may have two (or more) enemies which are even more severe enemies of each other. This is the principle behind the commonly expressed half-truth that grass benefits by being grazed (or mown). Cattle eat grass, and might therefore be thought of as enemies of grass. But grasses also have other enemies in the plant world, competitive weeds, which, if allowed to grow unchecked, might turn out to be even more severe enemies of grasses than cattle. Grasses suffer somewhat from being eaten by cattle, but the competitive weeds suffer even more. *(Grasses are well equipped to prosper in a heavily grazed field. Because their thin leaves grow from the base rather than the tip, they are unharmed by the nibbling of herbivores. Grasses are also uncommonly good at reproducing vegetatively by sprouting clones from their ever-extending roots; perennial grasses are thus able to spread even if grazing prevents them from setting seed.)*

The Red Queen Effect

The kernel of the arms-race idea is that both sides in the arms race are improving from their own point of view, while simultaneously making life more difficult for the other side in the arms race. There is no particular reason (or at least none in anything that we have discussed so far) to expect either side in the arms race to become steadily more successful or less successful than the other. In fact the arms-race idea, in its purest form, suggests that there should be absolutely zero progress in the success rate on both sides of the arms race, while there is very definite progress in the equipment for success on both sides. Predators become better equipped for killing, but at the same time prey become better equipped to avoid being killed, so the net result is no change in the rate of successful killings.

The principle of zero change in success rate, no matter how great the evolutionary progress in equipment, has been given the memorable name of the 'Red Queen effect' by the American biologist Leigh van Valen. In *Through the Looking Glass,* you will remember, the Red Queen seized Alice by the hand and dragged her, faster and faster, on a frenzied run through the countryside, but no matter how fast they ran they always stayed in the same place. Alice was understandably puzzled, saying, 'Well in our country you'd generally get to somewhere else—if you ran very fast for a long time as we've been doing.' 'A slow sort of country!' said the Queen. 'Now, here, you see, it takes all the running you can do, to keep in the same place. If you want to get somewhere else, you must run at least twice as fast as that!'

The Red Queen label is amusing, but it can be misleading if taken (as it sometimes is) to mean something mathematically precise, literally zero

relative progress. Another misleading feature is that in the Alice story the Red Queen's statement is genuinely paradoxical, irreconcilable with common sense in the real physical world. But van Valen's evolutionary Red Queen effect is not paradoxical at all. It is entirely in accordance with common sense, so long as common sense is intelligently applied. If not paradoxical, however, arms races can give rise to situations that strike the economically minded human as wasteful.

Why, for instance, are trees in forests so tall? The short answer is that all the other trees are tall, so no one tree can afford not to be. It would be overshadowed if it did. This is essentially the truth, but it offends the economically minded human. It seems so pointless, so wasteful. When all the trees are the full height of the canopy, all are approximately equally exposed to the sun, and none could afford to be any shorter. But if only they were all shorter, if only there could be some sort of trade-union agreement to lower the recognized height of the canopy in forests, all the trees would benefit. They would be competing with each other in the canopy for exactly the same sunlight, but they would all have 'paid' much smaller growing costs to get into the canopy. The total economy of the forest would benefit, and so would every individual tree.

Unfortunately, natural selection doesn't care about total economies, and it has no room for cartels and agreements. There has been an arms race in which forest trees became larger as the generations went by. At every stage of the arms race there was no intrinsic benefit in being tall for its own sake. At every stage of the arms race the only point in being tall was to be relatively taller than neighbouring trees.

Successive generations of trees got taller and taller, but at the end they might better, in one sense, have stayed where they started. Here, then, is the connection with Alice and the Red Queen, but you can see that in the case of the trees it is not really paradoxical. It is generally characteristic of arms races, including human ones, that although all would be better off if none of them escalated, so long as one of them escalates none can afford not to.

Asymmetric Arms Races and Runaway Evolution

The tree story allows me to introduce an important general distinction between two kinds of arms race, called symmetric and asymmetric arms races. A symmetric arms race is one between competitors trying to do roughly the same thing as each other. The arms race between forest trees struggling to reach the light is an example. The different species of trees are not all making their livings in exactly the same way, but as far as the particular race we are talking about is concerned—the race for the sunlight above the canopy—they are competitors for the same resource. They are taking part in an arms race in which success on one side is felt by the other side as failure. And it is a symmetric arms race because the nature of the success and failure on the two sides is the same: attainment of sunlight and being overshadowed, respectively.

The arms race between cheetahs and gazelles, however, is asymmetric. It is a true arms race in which success on either side is felt as failure by the other side, but the nature of the success and failure on the two sides is very different. The two sides are 'trying' to do very different things. Cheetahs are trying to eat gazelles. Gazelles are not trying to eat cheetahs, they are trying to avoid being eaten by cheetahs. From an evolutionary point of view asymmetric arms races are more interesting, since they are more likely to generate highly complex weapons systems. We can see why this is by taking examples from human weapons technology.

I could use the USA and the USSR as examples, but there is really no need to mention specific nations. Weapons manufactured by companies in any of the advanced industrial countries may end up being bought by any of a wide variety of nations. The existence of a successful offensive weapon, such as the Exocet type of surface skimming missile, tends to 'invite' the invention of an effective counter, for instance a radio jamming device to 'confuse' the control system of the missile. The counter is more likely than not to be manufactured by the same country, even by the same company! No company, after all, is better equipped to design a jamming device for a particular missile than the company that made the missile in the first place. There is nothing inherently improbable about the same company producing both and selling them to opposite sides in a war. I am cynical enough to suspect that it probably happens, and it vividly illustrates the point about equipment improving while its net effectiveness stands still (and its costs increase).

So far I have discussed the example of the missile and its specific antidote without stressing the evolutionary, progressive aspect, which is, after all, the main reason for bringing it into this chapter. The point here is that not only does the present design of a missile invite, or call forth, a suitable antidote, say a radio jamming device. The antimissile device, in its turn, invites an improvement in the design of the missile, an improvement that specifically counters the antidote, an anti-anti-missile device. It is almost as though each improvement in the missile stimulates the next improvement in itself, via its effect on the antidote. Improvement in equipment feeds on itself. This is a recipe for explosive, runaway evolution.

At the end of some years of this ding-dong invention and counter-invention, the current version of both the missile and its antidote will have attained a very high degree of sophistication. Yet at the same time— here is the Red Queen effect again—there is no general reason for expecting either side in the arms race to be any more successful at doing its job than it was at the beginning of the arms race. There has been progress in design, but no progress in accomplishment, specifically because there has been equal progress in design on both sides of the arms race. Indeed, it is precisely because there has been approximately equal progress on both sides that there has been so much progress in the level of sophistication of design. If one side, say the antimissile jamming device, pulled too far ahead in the design race, the other side, the missile in this case, would simply cease to be used and manufactured: it would go 'extinct'. Far from being paradoxical like Alice's original example, the Red Queen effect in its arms-race context turns out to be fundamental to the very idea of progressive advancement.

In the living world too, we shall expect to find complex and sophisticated design wherever we are dealing with the end-products of a long, asymmetric arms race in which advances on one side have always been matched, on a one-to-one, point-for-point basis, by equally successful antidotes (as opposed to competitors) on the other. This is conspicuously true of the arms races between predators and their prey, and, perhaps even more, of arms races between parasites and hosts. The electronic and acoustic weapons of bats have all the finely tuned sophistication that we expect from the end-products of a long arms race. Not surprisingly, we can trace this same arms race on the other side. The insects that bats prey upon have a comparable battery of sophisticated electronic and acoustic gear. Some moths even emit bat-like (ultra-) sounds that seem to put the bats off.

Almost all animals are either in danger of being eaten by other animals or in danger of failing to eat other animals, and an enormous number of detailed facts about animals makes sense only when we remember that they are the end-products of long and bitter arms races.

How Do Arms Races End?

How do arms races end? Sometimes they may end with one side going extinct, in which case the other presumably stops evolving in that particular progressive direction, and indeed it will probably even 'regress' for economic reasons. In other cases, economic pressures may impose a stable halt to an arms race, stable even though one side in the race is, in a sense, permanently ahead. Take running speed, for instance. There must be an ultimate limit to the speed at which a cheetah or a gazelle can run, a limit imposed by the laws of physics. But neither cheetahs nor gazelles have reached that limit. Both have pushed up against a lower limit which is, I believe, economic in character.

High-speed technology is not cheap. It demands long leg bones, powerful muscles, capacious lungs. These things can be had by any animal that really needs to run fast, but they must be bought. They are bought at a steeply increasing price. The price is measured as what economists call 'opportunity cost'. The opportunity cost of something is measured as the sum of all the other things that you have to forgo in order to have that something. The cost of sending a child to a private, fee-paying school is

all the things that you can't afford to buy as a result. The price, to a cheetah, of growing larger leg muscles is all the other things that the cheetah could have done with the materials and energy used to make the leg muscles, for instance make more milk for cubs.

There is no suggestion, of course, that cheetahs do cost-accounting sums in their heads! It is all done automatically by ordinary natural selection. A rival cheetah that doesn't have such big leg muscles may not run quite so fast, but it has resources to spare for making an extra lot of milk and therefore perhaps rearing another cub. More cubs will be reared by cheetahs whose genes equip them with the optimum compromise between running speed, milk production and all the other calls on their budget. When both cheetahs and gazelles reach the maximum running speed that they can 'afford', in their own internal economies, the arms race between them will come to an end.

Their respective economic stopping points may not leave them exactly equally matched. Prey animals may end up spending relatively more of their budget on defensive weaponry than predators do on offensive weaponry. One reason for this is summarized in the Aesopian moral: The rabbit runs faster than the fox, because the rabbit is running for his life, while the fox is only running for his dinner. *(This idea, developed by Richard Dawkins and coauthor John R. Krebs, is now called the Life/ Dinner Principle.)*

We are unlikely to witness arms races in dynamic progress, because they are unlikely to be running at any particular 'moment' of geological time, such as our time. But the animals that are to be seen in our time can be interpreted as the end-products of an arms race that was run in the past. Arms races have an inherently unstable 'runaway' feel to them. They career off into the future in a way that is, in one sense, pointless and futile, in another sense progressive and endlessly fascinating to us, the observers.

Negative and Positive Feedbacks in Biological Processes

[The runaway] 'positive feedback' [of evolutionary arms races] is best understood by comparison with its opposite, negative feedback. Negative feedback is the basis of most automatic control and regulation, and one of its neatest and best-known examples is the Watt Steam governor. A useful engine should deliver rotational power at a constant rate, the right rate for the job in hand, milling, weaving, pumping or whatever it happens to be. Before Watt, the problem was that the rate of turning depended upon the steam pressure. Stoke the boiler and you speed up the engine, not a satisfactory state of affairs for a mill or loom that requires uniform drive for its machines. Watt's governor was an automatic valve regulating the flow of steam to the piston.

The clever trick was to link the valve to the rotary motion produced by the engine, in such a way that the faster the engine ran the more the valve shut down the steam. Conversely, when the engine was running slowly, the valve opened up. Therefore an engine going too slowly soon speeded up, and an engine going too fast soon slowed down. The precise means by which the governor measured the speed was simple but effective, and the principle is still used today. A pair of balls on hinged arms spin round, driven by the engine. When they are spinning fast, the balls rise up on their hinges, by centrifugal force. When they are spinning slowly, they hang down. The hinged arms are directly linked to the steam throttle. With suitable fine-tuning, the Watt governor can keep a steam engine turning at an almost constant rate, in the face of considerable fluctuations in the firebox.

The underlying principle of the Watt governor is negative feedback. The output of the engine (rotary motion in this case) is fed back into the engine (via the steam valve). The feedback is negative because high output (fast rotation of the balls) has a negative effect upon the input (steam supply). Conversely, low output (slow rotation of the balls) boosts the input (of steam), again reversing the sign.

Engineers have found it fruitful to unite a wide variety of processes under the single heading of negative feedback, and another wide variety under the heading of positive feedback. The analogies are fruitful not just in some vague qualitative sense, but because all the processes share the same underlying mathematics. Biologists studying such phenomena as temperature control in the body, and the satiation mechanisms that prevent overeating, have found it helpful to borrow the mathematics of negative feedback from engineers.

Positive-feedback systems are used less than negative feedback, both by engineers and living bodies. The reason engineers and living bodies make more use of negative than positive-feedback systems is, of course, that controlled regulation near an optimum is useful. Unstable runaway processes, far from being useful, can be downright dangerous. In chemistry, the typical positive feedback process is an explosion, and we commonly use the word explosive as a description of any runaway process. The importance of positive feedbacks in international affairs is implicitly recognized in the jargon word 'escalation': when we say that the Middle East is a 'powder keg', and when we identify 'flashpoints'.

There are some features of living organisms that look as though they are the end-products of something like an explosive, positive-feedback-driven runaway process of evolution. In a mild way, arms races are examples of this, but the really spectacular examples are to be found in organs of sexual advertisement.

Sexual Selection

Try to persuade yourself, as they tried to persuade me when I was an undergraduate, that the peacock's fan is a mundanely functional organ like a tooth or a kidney, fashioned by natural selection to do no more than the utilitarian job of labelling the bird, unambiguously, as a member of this species and not that. They never persuaded me, and I doubt if you can be persuaded either. For me the peacock's fan has the unmistakable stamp of positive feedback. It is clearly the product of some kind of uncontrolled, unstable explosion that took place in evolutionary time. So thought Darwin in his theory of sexual selection.

Darwin, although he laid his main stress on survival and the struggle for existence, recognized that existence and survival were only means to an end. That end was reproduction. A pheasant may live to a ripe old age, but if it does not reproduce it will not pass its attributes on. Selection will favour qualities that make an animal successful at reproducing, and survival is only part of the battle to reproduce.

In other parts of the battle, success goes to those that are most attractive to the opposite sex. Darwin saw that, if a male pheasant or peacock or bird of paradise buys sexual attractiveness, even at the cost of its own life, it may still pass on its sexually attractive qualities through highly successful procreation before its death. He realized that the fan of a peacock must be a handicap to its possessor as far as survival is concerned, and he suggested that this was more than outweighed by the increased sexual attractiveness that it gave the male. With his fondness for the analogy with domestication, Darwin compared the hen to a human breeder directing the course of evolution of domestic animals along the lines of aesthetic whims.

Darwin simply accepted female whims as given. Their existence was an axiom of his theory of sexual selection, a prior assumption rather than something to be explained in its own right. Partly for this reason his theory of sexual selection fell into disrepute, until it was rescued by R. A. Fisher in 1930. Unfortunately, many biologists either ignored or misunderstood Fisher. The objection raised by Julian Huxley and others was that female whims were not legitimate foundations for a truly scientific theory. But Fisher rescued the theory of sexual selection, by treating female preference as a legitimate object of natural selection in its own right, no less than male tails.

Bird of paradise, male

Female preference is a manifestation of the female nervous system. The female nervous system develops under the influence of her genes, and its attributes are therefore likely to have been influenced by selection over past generations. Where others had thought of male ornaments evolving under the influence of static female preference, Fisher thought in terms of female preference evolving dynamically in step with male ornament.

It may seem rather sexist to assume that it is the females that would choose their mates, rather than the other way round. Actually, there are good theoretical reasons for expecting it to be this way round. What, at bottom, defines a female? We as mammals see the sexes defined by whole syndromes of characteristics—possession of a penis, bearing of the young, suckling by means of special milk glands, certain chromosomal features, and so on. These criteria for judging the sex of an individual are all very well for mammals but, for animals and plants generally, they are no more reliable than is the tendency to wear trousers as a criterion for judging human sex. In frogs, for instance, neither sex has a penis.

Perhaps, then, the words male and female have no general meaning. However, there is one fundamental feature of the sexes which can be used to label males as males, and females as females, throughout animals and plants. This is that the sex cells or 'gametes' of males are much smaller and more numerous than the gametes of females. The difference is especially pronounced in reptiles and in birds, where a single egg cell is big enough and nutritious enough to feed a developing baby for several weeks. Even in humans, where the egg is microscopic, it is still many times larger than the sperm.

Because a male produces many millions of sperms to every egg produced by a female, sperms heavily outnumber eggs in the population. Any given egg is therefore much more likely to enter into sexual fusion than any given sperm is. Eggs are a relatively valuable resource, and therefore a female does not need to be so sexually attractive as a male does in order to ensure that her eggs are fertilized. A male is perfectly capable of siring all the children born to a large population of females. Even if a male has a very short life because his gaudy tail attracts predators, or gets tangled in the bushes, he may have fathered a very large number of children before he dies. An unattractive or drab male may live even as long as a female, but he has few children, and his genes are not passed on. What shall it profit a male if he shall gain the whole world, and lose his immortal genes?

Testing Female Choice

Females have no difficulty in finding mates, and are in a position to be choosy. How should we go about looking for evidence? What methods might be used? A promising approach was made by Malte Andersson, from Sweden. He worked on the long-tailed widow bird, and he studied it in its natural surroundings in Kenya.

Andersson caught thirty-six male widow birds, and divided them into nine groups of four. Each group of four was treated alike. One member of each group of four (scrupulously chosen at random to avoid any unconscious bias) had its tail feathers trimmed to 14 centimeters (about 5.5 inches). The portion removed was stuck, with quick-setting superglue, to the end of the tail of the second member of the group of four. So, the first one had an artificially shortened tail, the second one an artificially lengthened tail. The third bird was left with his tail untouched, for comparison. The fourth bird was also left with his tail the same length, but it wasn't untouched. Instead, the ends of the feathers were cut off and then glued back on again. This might seem a pointless exercise, but it is a good example of how careful you have to be in designing experiments. It could have been that the fact of having his tail feathers manipulated, or the fact of being caught and handled by a human affected a bird, rather than the actual change in length itself. Group 4 was a 'control' for such effects.

The idea was to compare the mating success of each bird with its differently treated colleagues in its own group of four. After being treated in one of the four ways, every male was allowed to take up its former residence on its own territory. Here it resumed its normal business of trying to attract females into its territory, there to mate, build a nest and lay eggs. The question was, which member of each group of four would have the most success in pulling in females? Andersson measured this, not by literally watching females, but by waiting and then counting the number of nests containing eggs in each male's territory. What he found was that males with artificially elongated tails attracted nearly four times as many females as males with artificially shortened tails. Those with tails of normal, natural length had intermediate success.

The results were analysed statistically, in case they had resulted from chance alone. The conclusion was that if attracting females were the only criterion, males would be better off with longer tails than they actually have. In other words, sexual selection is constantly pulling tails (in an evolutionary sense) in the direction of getting longer. The fact that real tails are shorter than females would prefer suggests that there must be some other selection pressure keeping them shorter. This is 'utilitarian' selection. Presumably males with especially long tails are more likely to die than males with average tails. Unfortunately, Andersson did not have time to follow the subsequent fates of his doctored males.

Peacock fans, and widow bird and bird of paradise tails, in their gaudy extravagance, are very plausibly seen as end-products of explosive, spiralling evolution by positive feedback. Fisher and his modern successors have shown us how this might have come about. Is this idea essentially tied to sexual selection, or can we find convincing analogies in other kinds of evolution?

It is worth asking this question, if only because there are aspects of our own evolution that have more than a suggestion of the explosive about them, notably the extremely rapid swelling of our brains during the last few million years. It has been suggested that this is due to sexual selection itself, braininess being a sexually desirable character (or some manifestation of braininess, such as ability to remember the steps of a long and complicated ritual dance).

Richard Dawkins is an eloquent spokesperson for a research tradition and a worldview that began with Charles Darwin. Whether the metaphor is wedges, arms races, or Tennyson's "red in tooth and claw" the role of strife is central to this perspective: the strife between members of the same species competing for limited resources or mates, the strife between species engaged in symmetric or asymmetric arms races, even the kinds of strife in which there is no "enemy"—say, the struggle to survive in a parched or a bone-chilling land.

———

Are God and Nature then at strife
   That Nature lends such evil dreams?
   So careful of the type she seems,
So careless of the single life . . .

'So careful of the type?' but no.
   From scarped cliff and quarried stone
   She cries, 'A thousand types are gone:
I care for nothing, all shall go.'

—Alfred, Lord Tennyson

In the previous chapter, Charles Darwin and (even more so) Richard Dawkins pointed to competitive strife as the primary tool of biological evolution. "Survival of the fittest" is the leitmotif of this scientific outlook in which wedges and arms races are useful metaphors. While few scientists today would deny that competition plays a vital role in evolution, some believe that integrative processes are equally or even more important. From this perspective, evolution is driven by "survival of those that fit" into the web of ecological relationships.

Richard Dawkins introduced the Watt steam governor in order to build a contrast with the dynamics of arms races and sexual selection. Dawkins tips his hat to the essential role played by negative feedback controls within the bodies of individual organisms, but he credits positive feedback between embattled species or sexes as the sculptor of more and more sophisticated (or merely flamboyant) equipment within the tree of life. Negative feedback and the governor analogy are viewed by some, however, as tool and metaphor of biological evolution in their own right.

Gregory Bateson is one proponent of an integrative view of evolution. Until his death in 1980 Bateson roamed across intellectual boundaries and made significant contributions to psychology (notably, the double bind theory of schizophrenia) and to cybernetics, which is the study of electrical, mechanical, and biological systems that exhibit the kind of circular causal loops by which steam engines equipped with Watt governors self-regulate.

Here, Bateson's observations about circular causality, negative feedback, and evolution are drawn from two books. Passages from Steps to an Ecology of Mind *by Gregory Bateson (copyright 1972 by Chandler Publishing Company) are reprinted courtesy of Jason Aronson Inc., Publishers. Passages from* Mind and Nature *by Gregory Bateson (copyright 1979 by Gregory Bateson) are reprinted by permission of Dutton, an imprint of New American Library, a division of Penguin Books USA, Inc.*

The Governor Analogy as an Evolutionary Mechanism

Gregory Bateson   In the theory of history, Marxian philosophy, following Tolstoi, insists that the great men who have been the historic nuclei for profound social change or invention are, in a certain sense, irrelevant to the changes they precipitated. It is argued, for example, that in 1859, the occidental world was ready and ripe (perhaps over-ripe) to create and receive a theory of evolution that could reflect and justify the ethics of the Industrial Revolution. From that point of view, Charles Darwin himself could be made to appear unimportant. If he had not put out his theory, somebody else would have put out a similar theory within the next five years. Indeed, the parallelism between Alfred Russel Wallace's theory and that of Darwin would seem at first sight to support this view.

The story is worth repeating. Wallace was a young naturalist who, in 1856 (three years before the publication of Darwin's *Origin*), while in the rain forests of Ternate, Indonesia, had an attack of malaria and, following delirium, a psychedelic experience in which he discovered the principle of natural selection. He wrote this out in a long letter to Darwin. In this letter he explained his discovery in the following words:

The action of this principle is exactly like that of the centrifugal governor of the steam engine, which checks and corrects any irregularities almost before they become evident; and in like manner no unbalanced deficiency in the animal kingdom can ever reach any conspicuous magnitude because it would make itself felt at the very first step, by rendering existence difficult and extinction almost sure to follow.

It is, I claim, nonsense to say that it does not matter which individual man acted as the nucleus for the change. It is precisely this that makes history unpredictable into the future. The Marxian error is a simple blunder in logical typing, a confusion of individual with class. If it had been Wallace instead of Darwin, we would have had a very different theory of evolution today. Wallace, in fact, proposed the first cybernetic model. The whole cybernetics movement might have occurred a hundred years earlier as a result of Wallace's comparison between the steam engine with a governor and the process of natural selection. Or perhaps the big theoretical step might have occurred in France and evolved from the ideas of Claude Bernard who, in the late nineteenth century, discovered what later came to be called the homeostasis of the body. He observed that the 'milieu interne'—the internal environment—was balanced or self-correcting.

When the phenomena of the universe are seen as linked together by cause-and-effect and energy transfer, the resulting picture is of complexly branching and interconnecting chains of causation. In certain regions of this universe (notably organisms in environments, ecosystems, thermostats, steam engines with governors, societies, computers, and the like), these chains of causation form circuits which are closed in the sense that causal interconnection can be traced around the circuit and back through whatever position was (arbitrarily) chosen as the starting point of the description. In such a circuit, evidently, events at any position in the circuit may be expected to have effect at all positions on the circuit at later times.

In a balanced ecological system whose underpinnings are of this nature, it is very clear that any monkeying with the system is likely to disrupt the equilibrium. Some plant will become a weed, some creatures will be exterminated, and the system as a balanced system is likely to fall to pieces. What is true of the species that live together in a wood is also true of the groupings and sorts of people in a society, who are similarly in an uneasy balance of dependency and competition. And the same truth holds right inside you, where there is an uneasy physiological competition and mutual dependency among the organs, tissues, cells, and so on.

The Mental Character of Cybernetic Systems

Gregory Bateson thus believed that Wallace's metaphor for natural selection would have promoted a biology more akin to his own vision of a cybernetic, ecological, and interlinked world. Nevertheless, he espies in the governor analogy a serious flaw:

Extraordinary advances have been made in our knowledge of what sort of thing the environment is, what sort of thing an organism is, and, especially, what sort of thing a mind is. These advances have come out of cybernetics, systems theory, information theory, and related sciences. Most relevant in the present context, we know that no part of an internally interactive system can have unilateral control over the remainder or over any other part. Even in very simple self-corrective systems, this holistic character is evident. In the steam engine with a governor, the very word 'governor' is a misnomer if it be taken to mean that this part of the system has unilateral control. The governor is, essentially, a sense organ or transducer which receives a transform of the difference between the actual running speed of the engine and some ideal or preferred speed. This sense organ transforms these differences into differences in some efferent message, for example, to fuel supply or to a brake. The behavior of the governor is determined, in other words, by the behavior of the other parts of the system, and indirectly by its own behavior at a previous time.

The holistic and mental character of the system is most clearly demonstrated by this last fact, that the behavior of the governor (and, indeed, of every part of the causal circuit) is partially determined by its own previous behavior. There is thus a sort of determinative 'memory' in even the simplest cybernetic circuit. Even a human governor in a social system is bound by the same limitations. He is controlled by information from the system and must adapt his own actions to its time characteristics and to the effects of his own past action. Thus, in no system which shows mental characteristics can any part have unilateral control over the whole. In other words, the mental characteristics of the system are immanent, not in some part, but in the system as a whole.

Evolution [used to be thought of as] the history of how organisms learned more tricks for controlling the environment; and man had better tricks than any other creature. But that arrogant scientific philosophy is now obsolete, and in its place there is the discovery that man is only a part of larger systems and that the part can never control the whole. Goebbels thought that he could control public opinion in Germany with a vast communication system, and our own public relations men are perhaps liable to similar delusions. But in fact the would-be controller must always have his spies out to tell him what the people are saying about his propaganda. He is therefore in the position of being responsive to what they are saying. Therefore he cannot have a simple lineal control. We do not live in the sort of universe in which simple lineal control is possible. Life is not like that.

The Ladder Metaphor versus Immanence of Mind

In mid-eighteenth century the biological world looked like this: there was a supreme mind at the top of the ladder, which was the basic explanation of everything downwards from that—the supreme mind being, in Christianity, God. The ladder of explanation went downwards deductively from the Supreme to man to the apes, and so on, down to the infusoria *(an archaic term for the one-celled organisms that became visible under the earliest microscopes)*. This hierarchy was a set of deductive steps from the most perfect to the most crude or simple. And it was rigid. It was assumed that every species was unchanging.

Lamarck, probably the greatest biologist in history, turned that ladder of explanation upside down. He was the man who said it starts with the infusoria and that there were changes leading up to man. His turning the taxonomy upside down is one of the most astonishing feats that has ever occurred. It was the equivalent in biology of the Copernican revolution in astronomy.

The logical outcome of turning the taxonomy upside down was that the study of evolution might provide an explanation of mind. Up to Lamarck, mind was the explanation of the biological world. But, hey presto, the question now arose: Is the biological world the explanation of mind? That which was the explanation now became that which was to be explained. About three quarters of Lamarck's *Philosophie Zoologique* (1809) is an attempt, very crude, to build a comparative psychology. He achieved and formulated a number of very modern ideas: that you cannot attribute to any creature psychological capacities for which it has no organs; that mental process must always have physical representation; and that the complexity of the nervous system is related to the complexity of mind.

There the matter rested for 150 years, mainly because evolutionary theory was taken over, not by a Catholic heresy but by a Protestant heresy,

in the mid-nineteenth century. Darwin's opponents, you may remember, were not Aristotle and Aquinas, who had some sophistication, but fundamentalist Christians whose sophistication stopped with the first chapter of Genesis. The question of the nature of mind was something which the nineteenth-century evolutionists tried to exclude from their theories, and the matter did not come up again for serious consideration until after World War II.

In World War II *(when cybernetics arose)* it was discovered what sort of complexity entails mind. And, since that discovery, we know that wherever in the Universe we encounter that sort of complexity, we are dealing with mental phenomena. It's as materialistic as that.

———

Take your well-disciplined strengths
and stretch them between two
opposing poles. Because inside human beings
is where God learns.

—*Rainer Maria Rilke*

———

Imperfections in Science and Nature

Gregory Bateson perceived mental processes or mind in a rather expansive group of systems, all sharing a circular form of causation. Ecological relationships and evolutionary change triggered by feedbacks were, to Bateson, representative of something he was quite comfortable to call "mind." Mind, moreover, emerges only within these special systems; thus, mind is immanent, not transcendent. Mind does not stand godlike outside the physical realm; mind comes into being in tandem with and within a universe growing more complex.

Added to the earlier chapter that gave voice to Charles Darwin, Bateson's attention here to Alfred Russel Wallace and Jean-Baptiste Lamarck fills out the triumvirate of naturalists publishing in the nineteenth century who are widely credited with launching the scientific basis of biological evolution. All of them, however, were fundamentally wrong in some of their formulations. Lamarck, sadly, is most remembered for his faulty theory that physical qualities acquired during one's life pass through to offspring. In his later years Darwin, too, leaned toward this Lamarckian view, and his own theory of inheritance by "pangenesis" was equally fanciful. An aging Wallace gravitated to a dogged divorce of soul from body; his view of the relationship between mind and matter was a retreat from Lamarck's bold reversal of the ladder metaphor.

In hindsight, no scientist or scientific theory stands untarnished. Indeed, a modern view of science as a whole is that it is a quest not for truth but for a closer and closer approximation of a truth that can never be completely and assuredly known. Then too, the popular notion that biological evolution yields perfection—in bodily features, sensory organs, adaptive behaviors—is itself a myth, one grounded in the ideals of eighteenth-century naturalists and clergy who employed the "argument of design" to meld science and the Christian faith into an all-encompassing worldview. Darwin stood this theological argument on its head. He pointed to the plethora of imperfections in the living world and thereby challenged the belief that species had arisen by fiat of a divine creator.

In the excerpts that follow, François Jacob (a French biologist who shared a Nobel Prize with Jacques Monod and André Lwoff) explores the glorious imperfections of the living world. To Jacob, natural processes are more like a tinkerer than an engineer. Honed by natural selection, the mishmash of parts and purposes is nevertheless integrated into an organism that, at least for a time, can make its way in an unforgiving world.

Selections are drawn from The Possible and the Actual *by François Jacob (copyright 1982 by François Jacob). Permission is granted by the author and by Pantheon Books, a division of Random House, Inc. As with all other excerpts in this anthology, citations in the original text have been omitted.*

Evolution as Tinkering

François Jacob

The action of natural selection has often been compared to that of an engineer. This comparison, however, does not seem suitable. First, in contrast to what occurs during evolution, the engineer works according to a preconceived plan. Second, an engineer who prepares a new structure does not necessarily work from older ones. The electric bulb does not derive from the candle, nor does the jet engine descend from the internal combustion engine. To produce something new, the engineer has at his disposal original blueprints drawn for that particular occasion, materials and machines specially prepared for that task. Finally, the objects thus produced de novo by the engineer, at least by the good engineer, reach the level of perfection made possible by the technology of the time.

In contrast, evolution is far from perfection, as was repeatedly stressed by Darwin, who had to fight against the argument from perfect creation. In the *Origin of Species,* Darwin emphasizes over and over again the structural and functional imperfections of the living world. He always points out the oddities, the strange solutions that a reasonable God would never have used.

In contrast to the engineer, evolution does not produce innovations from scratch. It works on what already exists, either transforming a system to give it a new function or combining several systems to produce a more complex one. Natural selection has no analogy with any aspect of human behavior. If one wanted to use a comparison, however, one would

Flounder larvae look like normal fish, but one eye gradually moves across the top of the head and down the other side. As adults, these bottom-dwelling flatfish swim and rest on one side.

have to say that this process resembles not engineering but tinkering, *bricolage* we say in French.

While the engineer's work relies on his having the raw materials and the tools that exactly fit his project, the tinkerer manages with odds and ends. Often without even knowing what he is going to produce, he uses whatever he finds around him, old cardboards, pieces of string, fragments of wood or metal, to make some kind of workable object. As pointed out by Claude Levi-Strauss, none of the materials at the tinkerer's disposal has a precise and definite function. Each can be used in different ways. What the tinkerer ultimately produces is often related to no special project. It merely results from a series of contingent events, from all the opportunities he has had to enrich his stock with leftovers. In contrast with the engineer's tools, those of the tinkerer cannot be defined by a project. What can be said about any of these objects is just that "it could be of some use." For what? That depends on the circumstances.

In some respects, the evolutionary derivation of living organisms resembles this mode of operation. In many instances, and without any well-defined long-term project, the tinkerer picks up an object which happens to be in his stock and gives it an unexpected function. Out of an old car wheel, he will make a fan; from a broken table, a parasol. This process is not very different from what evolution performs when it turns a leg into a wing, or a part of a jaw into a piece of ear.

Evolution proceeds like a tinkerer who, during millions of years, has slowly modified his products, retouching, cutting, lengthening, using all opportunities to transform and create. The formation of a lung in terrestrial vertebrates, as described by Ernst Mayr, provides a clear example of this process. Lung development started in certain freshwater fishes living in stagnant pools lacking oxygen. They adopted the habit of swallowing air and absorbing oxygen throughout the walls of the esophagus. Under such conditions, enlargement of the surface area of the esophagus conferred a selective advantage. Diverticula of the esophagus appeared and, under continuous selective pressure, enlarged into lungs. Further evolution of the lung was merely an elaboration of the same theme: enlarging the surface for oxygen uptake and vascularization. Making a lung with a piece of esophagus sounds very much like making a skirt with a piece of Granny's curtain.

When different engineers tackle the same problem, they are likely to end up with very nearly the same solution: all cars look alike, as do all cameras and all fountain pens. In contrast, different tinkerers interested in the same problem will reach different solutions, depending on the opportunities available to each of them. This variety of solutions also applies to the products of evolution, as is shown, for instance, by the diversity of eyes found throughout the living world. The possession of light receptors confers a great advantage under a variety of conditions. During evolution, many types of eyes appeared, based on at least three different principles: the lens, the pinhole, and multiple holes. The most sophisticated ones, like ours, are lens-based eyes, which provide information not only on the intensity of incoming light but also on the objects the light comes from, on their shape, color, position, motion, speed, distance, and so forth. Such sophisticated structures are necessarily complex.

One might suppose, therefore, that there is just one way of producing such a structure. But this is not the case. Eyes with lenses have appeared in molluscs and in vertebrates. Nothing looks so much like our eye as the octopus eye. Yet it did not evolve the same way. In vertebrates, the photoreceptor cells of the retina point away from light while in molluscs they point toward light. Among the many solutions found to the problem of photoreceptors, these two are similar but not identical. In each case, natural selection did what it could with the materials at its disposal.

A Final Example of Tinkering: The Human Brain

Finally, in contrast with the engineer, the tinkerer who wants to refine his work will often add new structures to the old ones rather than replace them. This procedure is also frequently observed with evolution, as exemplified by the development of the brain in mammals. This development

Queen conch (the two
eyes have lenses)

was not as integrated a process as, for instance, the transformation of a leg into a wing. It involved the addition, to the old rhinencephalon of lower vertebrates, of a neocortex which rapidly, perhaps too rapidly, took a most important part in the evolutionary sequence leading to man.

Some neurobiologists, especially Paul McLean, consider that those two types of structures correspond to two types of functions and have not been completely coordinated or hierarchized. The recent one, the neocortex, controls intellectual, cognitive activities while the old one, derived from the rhinencephalon, controls emotional and visceral activities. The old structure, which in lower mammals was in total command, has in men been relegated to the department of emotions, and constitutes what McLean calls the "visceral brain." Perhaps because development is so prolonged and maturity so delayed in man, this center maintains strong connections with lower autonomic centers and continues to coordinate such fundamental drives as obtaining food, hunting for a sexual partner, or reacting to an enemy.

This evolutionary procedure—the formation of a dominant neocortex coupled with the persistence of a nervous and hormonal system partially, but not totally, under the rule of the neocortex—strongly recalls the way the tinkerer works. It is somewhat like adding a jet engine to an old horse cart. No wonder accidents occur.

It is hard to realize that the living world as we know it is just one among many possibilities; that its actual structure results from the history of the earth. Yet living organisms are historical structures: literally creations of history. They represent, not a perfect product of engineering, but a patchwork of odd sets pieced together when and where opportunities arose. For the opportunism of natural selection is not simply a matter of indifference to the structure and operation of its products. It reflects the very nature of a historical process, full of contingency.

———

Diving down the banked corals
I brush a polyp cluster with one bare arm,
Bruising its delicate loops and rows.
This coral boulder carries life at the edge
Blooming on the lately dead.
So, here is another animal
With surprisingly thin skin.
Using the same solutions, our various
Bodies seal the ocean from the flesh:
Our skins hold the line against the sea.

Others have different styles:
Boning up their surfaces,
As obviously as snails.

Still using Cambrian ingenuity
To suck calcium from the sea
and spin it out as shell—that trick
On which bone making trades.

We carry our boundaries
In ancient rank,
Inheriting not only ocean
Blood, but also bone.

—*Carrol B. Fleming*

———

Symbiosis

François Jacob uses the metaphor of tinkering, "a patchwork of odd sets pieced together," to stress the opportunistic character and exquisite imperfections of biological evolution. For Lynn Margulis, however, the tree of life is quite literally a patchwork—one branch grafted onto another, so to speak. In fact, the configuration of a tree below ground is, from her perspective, a better proxy for the geometry of evolution than the tree above. Key events in the history of life resemble more the upward merging of roots than the skyward bifurcation of branches.

Now at the University of Massachusetts, Amherst, Margulis's work in cell biology and her relentless promotion of the findings and ideas of the small group of researchers studying symbiosis have finally convinced the majority of scientists that once-independent organisms fused to form more and more complex beings. Margulis and paleontologist Mark McMenamin, an expert on the earliest fossils of multicellular organisms, recently teamed to produce an article on symbiosis for The Sciences *magazine. "Marriage of Convenience" appeared in the September–October 1990 issue. Excerpts are reprinted with permission of the authors and the magazine, which is produced by the New York Academy of Sciences, New York City.*

Lynn Margulis, Mark McMenamin

Symbiosis has shaped the features of many organisms. The great evergreen forests that spread across the northern latitudes would wither and die without the threads of symbiotic fungi that extract nutrients from rocks and soil and convey them to the tree roots. Termites would be no threat to houses, except that their guts contain myriad bacteria and other, larger creatures capable of digesting the cellulose in wood. The giant tube worms that live near hot springs on the ocean floor lack mouths; they take nourishment from symbiotic bacteria that live in their tissues, metabolizing energy-rich sulfide compounds carried out of the earth's crust by the springs. In these cases the union of two or more kinds of organism has yielded what is in essence a new organism.

But symbiosis may have had a still more profound role in evolution. It may have been critical to the emergence not just of specific groups of organisms but of fungi, plants, animals and all other life-forms made up of eukaryotic cells: cells that, unlike bacteria, contain a nucleus and many specialized subunits, or organelles. In the mid 1960s one of us (Margulis) pursued an explanation for strange genetic data—data suggesting the eukaryotic cell itself originated in a series of ancient symbiotic unions. By now it is widely accepted that two kinds of organelle were once free-living bacteria that became established within the confines of other bacteria.

The idea that cells owe some of their complexity to symbiotic microorganisms originated long before biologists had the tools needed to explore the notion. Eighty years ago the Russian biologist Konstantin Sergeivich Mereschkovsky began to suspect that the chloroplasts of plant cells—the green speckles that capture sunlight and produce sugars and oxygen through photosynthesis—are interlopers. Mereschkovsky realized that the organelles resemble blue-green photosynthetic bacteria, called cyanobacteria, more closely than they do any other structure within the cell. He also knew that chloroplasts reproduce on their own, independent of the cell's division cycle. They simply split, or fission, as ordinary bacteria do, but they do so in the confines of the cell. Mereschkovsky and his colleague Andrei Sergeivich Famintzyn, who tried to grow isolated chloroplasts, proposed that the organelles are actually cyanobacteria that took up residence in an early ancestor of plant cells and eventually lost their autonomy.

That scenario was ignored or flatly rejected. But decades later, in the 1960s, the electron microscope showed that chloroplasts contain an intricate stack of internal membranes, similar to the ones in cyanobacteria. Investigators examining chloroplasts also spotted ribosomes, the molecular factories in which proteins are assembled. Ribosomes are a hallmark of independent cells. Eventually DNA, the molecule of heredity, was found in chloroplasts. In its organization the DNA had much more in common with the DNA of certain cyanobacteria than with nuclear DNA. These findings were accepted as clear confirmation of the former independence of the chloroplasts.

Evidence for the bacterial origin of another set of organelles, the mitochondria, accumulated in much the same way. These rice-shaped subunits are the power stations of the eukaryotic cell, where molecules from food react with oxygen during aerobic respiration, yielding a substance called ATP. Like the electricity generated by a power plant, ATP made in the mitochondria is a convenient and portable energy source for use elsewhere. Mitochondria, like chloroplasts, resemble bacteria and reproduce on their own; their appearance and behavior led Ivan E. Wallin, an anatomist at the University of Colorado Medical School in Denver, to conclude in the 1920s that mitochondria too originated as bacterial intruders.

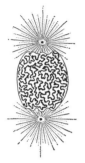

Wallin, who went on to propose a broader theory of the symbiotic origin of species, was spurned by biologists of the time. But in 1966 decisive evidence for the validity of his proposal about mitochondria came with the discovery that these organelles have their own DNA. Later, comparisons of mitochondrial DNA with DNA from various kinds of bacteria revealed parallels to certain purple nonsulfur bacteria—photosynthetic microorganisms that can also carry out aerobic respiration. Their ancestors are the most likely symbiotic precursors of mitochondria.

Fusion Evolution

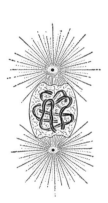

An image of piecemeal evolution emerges from these findings. The eukaryotic cell did not emerge from a single precursor cell—a bacterium of some kind—that gradually evolved more sophisticated features. Rather it arose from several organisms that interacted closely. Each precursor contributed an entire module of genes, which specified a distinctive set of biochemical abilities.

The kind of bacterium that accommodated the other symbionts in its interior may have been one similar to *Thermoplasma,* a tough microorganism living in acidic hot springs. Like all other bacteria, *Thermoplasma* has DNA that floats freely in the cytoplasm, the jellylike substance of the cell. (Eukaryotic DNA, in contrast, is bundled into dark rod-shaped structures, or chromosomes, which in turn are enclosed in the nuclear membrane.) Yet *Thermoplasma* and its relatives differ from other bacteria in one crucial way: their DNA is coated with a protein similar to the histones that form the scaffolding of chromosomes in eukaryotes. Histones are conspicuously absent from other kinds of bacteria. Some hardy bacterium resembling *Thermoplasma* may have been the ancestral cell, which acquired additional metabolic abilities wholesale by taking in other bacteria.

These microbial interactions took place at a critical juncture in the history of life. Before two billion years ago there was little oxygen in the atmosphere, but as photosynthetic bacteria (including the cyanobacterial precursors of the chloroplasts) spread, the concentration of this gas rose. Oxygen, a poison to most of the microorganisms that represented the universe of life at the time, spurred the evolution of respiration in a few species. When tough, *Thermoplasma*-like ancestral bacteria began housing bacteria capable of respiration, probably after surviving invasion by these aggressive microorganisms, they gained a way of removing any oxygen that penetrated their membranes and, in the long term, a new way of deriving energy. Equipped with the precursors of mitchondria, the new symbiotic complexes spread into environments neither component organism could have colonized.

Later the metabolic repertoire of some of these compound cells was enlarged still further. They fed on carbohydrate-rich photosynthetic bac-

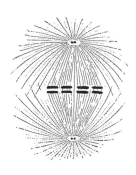

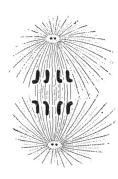

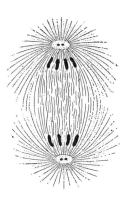

Dance of the chromosomes. *Left:* early stages of mitosis in the nucleus of a sea urchin egg. *Right:* later stages of mitosis as the chromosomes separate to form two new cells.

teria, but eventually some of the microbial prey resisted being digested. Surviving within the cell, the photosynthetic bacteria turned into dependent guests and ultimately into chloroplasts. The evolution of green algae (the precursors of the green plants) had begun.

This account is speculative, but the phenomena it describes—predatory relations in which microorganisms are the aggressors or the prey, followed by survival, coexistence and symbiosis—are seen in nature today. And one need not look hard to find the organisms that recently acquired new metabolic abilities when foreign microorganisms became incorporated into their own cells. Witness the tube worms, clams and mussels that form oases of life on the ocean floor, surviving on food synthesized from carbon dioxide in seawater by the sulfide-oxidizing bacteria that live in their tissues. Just as the bacterial precursors of mitochondria inadvertently protected their hosts from oxygen, the sulfide-oxidizing bacteria convert into a benign form the sulfide that would otherwise poison the animals.

Cell Motility through Symbiosis

What other feature of the eukaryotic cell might have evolved through symbiosis? Eukaryotic cells have a dynamic quality that bacteria cannot match. Some can change shape by extending tentacles or even broad skirts of membrane. Others use whiplike appendages to swim through a watery medium or sweep material across their own surfaces. Eukaryotic cells draw in food from their environment, shunting it around in membranous sacs. They rearrange their organelles individually and their cytoplasm in bulk. And during mitosis—the process of cell division characteristic of eukaryotes—they engage in a dance of the chromosomes. Typically the chromosomes line up at the center of the dividing cell, and the halves of each chromosome are drawn apart as two new cells form. (Bacterial cells simply duplicate their loop of DNA and then pinch apart.)

Bacteria lack all these forms of movement, though many bacteria can swim. They do so by means of flagella—minute rigid appendages that rotate like propellers. Eukaryotic cells such as sperm cells, the hair cells that sweep mucus up the lining of the throat and the cells of many protoctists (a kingdom that includes microscopic protists, algae, slime molds and other simple eukaryotes) wield much more complex appendages, flexible along their length, that lash and undulate. Given the differences between bacterial and eukaryotic flagella, it is better to revive a fifty-year-old term from the German and Russian literature and refer to the latter appendages as undulipodia (waving feet).

Undulipodia are just one expression of a motility system that extends throughout the eukaryotic cell, animating the cell interior as well as its exterior. We believe that this motility system, like chloroplasts and mitochondria, shows signs of having originated in formerly independent mi-

Mixotricha paradoxa is an intimate association of a wood-digesting large protist with four kinds of bacteria, including two types of spirochetes. Seven of the larger unidentified spirochetes can be seen, along with a population of treponemes that covers the protist's surface like fur. The small dots near the base of each treponeme are external symbiotic rod bacteria. The spherical objects tucked into the pleated sections of the ribbonlike endoplasmic reticulum are internal symbionts. The four threads at the top are the protist's own undulipodia.

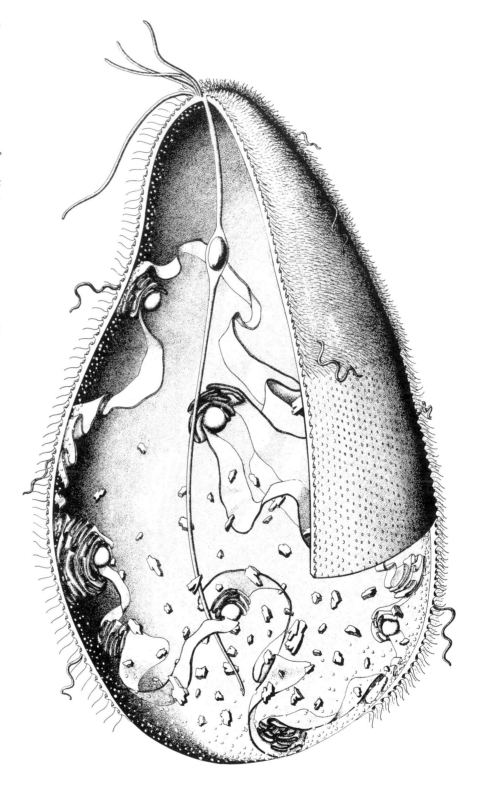

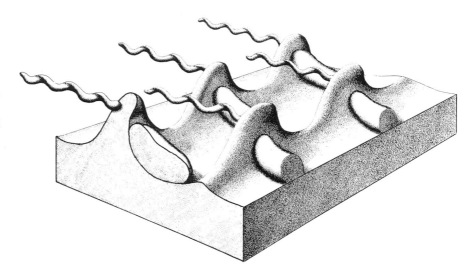

In a further enlargement of the surface of *Mixotricha,* the spiral form of the small treponeme is clearly visible, as are the rod bacteria that rest between raised structures of the protist's own surface. (Drawings by Christie Lyons)

croorganisms. The establishment of a symbiosis between these motility precursors and a *Thermoplasma*-like bacterium may have been the first step on the road to the protoctists, and ultimately to fungi, plants and animals.

Again there is a Russian precedent. In 1924 the biologist Boris Kozo-Polyansky proposed that undulipodia are relics of active, motile bacteria that once clung to an ancient protocell, acting as a kind of outboard motor. Kozo-Polyansky realized his proposal had broader implications. He saw that once the motile bacteria became an integral part of the protocell, they might also have influenced internal processes, including cell division.

Kozo-Polyansky's suggestion was unknown in the West until recently. Meanwhile, Margulis independently proposed the idea in a more modern form. In this view, early eukaryotic cells—immobile or slow-moving at best—gained the ability to move rapidly when they were joined by slender, motile confederates. The kinds of organisms most likely to have filled this role are spirochetes—common spiral-shaped bacteria, some kinds of which cause syphilis or Lyme disease. Present-day spirochetes not only are slender and fast-moving but also tend to associate with other cells, often grazing on their surfaces or even boring into their interiors.

These predatory relations can give way to symbiosis. One of the most vivid examples—what may be a replay of an event early in the history of life—comes from the hindgut of a termite that lives only near Darwin, Australia. Among the menagerie of organisms that help the insect digest the cellulose it eats is the protist *Mixotricha.* A large single-celled organism that ingests crumbs of wood and releases compounds the termite can digest, *Mixotricha* has a cluster of four undulipodia at one end. They serve only as rudders; *Mixotricha* gets all its forward impetus from spirochetes—hundreds of thousands of them, clustered as densely as hairs over its entire surface.

The flagella of most motile bacteria extend outside the cell wall, but spirochetes carry their flagella internally. As the flagella rotate, the entire length of the spirochete flexes back and forth, just as eukaryotic undulipodia do. Thus the spirochetes cloaking *Mixotricha* (the name means mixed-up hairs) are easy to mistake for undulipodia. They move in synchrony, like the oars of a galleon, because they are packed so closely.

Striking as this contemporary illustration of symbiotic spirochetes is, the ancestral association of spirochetes and *Thermoplasma*-like bacteria would have gone much further. The earlier symbiotic spirochetes must have gradually lost their metabolic self-sufficiency, since they could rely on the larger bacterium for food and protection against heat and acidity. They must have also lost genetic autonomy as genes were transferred from spirochete DNA to the histone-coated *Thermoplasma* DNA— the precursor of the nucleus.

Reenvisioning the Tree of Life

If the ability of eukaryotic cells to move and rearrange their contents turns out to be the legacy of symbiotic microorganisms, biologists will have gained deeper insight into how early cells crossed the evolutionary gulf separating bacteria and eukaryotes. The acquisition of motility, after all, must have preceded the advent of chloroplasts and mitochondria. All animal cells lack chloroplasts, and many obscure protists (*Mixotricha,* for example) lack even mitochondria. But nearly every nucleated organism has motility apparatus which accounts both for standard features of eukaryotic cells such as chromosome migration during mitosis and for so much of their wonderful diversity of shape and movement.

If the eukaryotic cell is viewed as a community of microorganisms, much of cell biology will be cast in a new light. One example is differentiation, the cellular specialization that goes on in many-celled organisms. The differentiation that yields a heart-muscle cell packed with mitochondria, a human sperm with its undulipodium or a chloroplast-laden photosynthesizing cell in a blade of grass might be viewed (as it was seventy years ago by Wallin) as the disproportionate growth of one or another of the microbial components of the nucleated cell. And if symbiosis gave rise to something as elaborate and unlikely as the eukaryotic cell, how many other key evolutionary advances may have come about through past symbiotic alliances?

Ordinary evolutionary change, incremental in nature, is hard put to account for some of the sudden advances in the fossil record. It also has trouble explaining how complex new structures and fine-tuned metabolic abilities could have arisen. An incremental step toward a new capability might handicap an organism by impairing an existing one. Only the full-fledged development, at the far side of some evolutionary barrier, might be viable. As existing symbioses—tube worms, termites, and the like—make clear, partnership with microorganisms provides a ready way of tunneling through such barriers. The larger organism gains all the necessary genes at a stroke, packaged in a tamed microbial invader or in undigested microbial food.

Over time the more familiar processes of evolution—mutation and selection—tend to eliminate many of the distinctions between symbiotic partners. Thus the branches on the tree of life do not always diverge. One branch can merge with another, and from these unions new limbs can grow, unlike anything seen before.

———

And these tend inward to me, and I tend outward to them,
And such as it is to be of these more or less I am,
And of these one and all I weave the song of myself.

—*Walt Whitman*

———

Synergism

Lynn Margulis and Mark McMenamin do not envision evolutionary change through symbiosis as inimical to the "red in tooth and claw" school of thought. In fact, they suspect that the paths to win-win symbioses may have been fraught with conflict—that mitochondria began as invasive parasites, that the hapless predecessors of chloroplasts were ingested as food.

McMenamin has elsewhere highlighted a similar coupling of conflict and cooperation (but in reverse) at a much later stage of biological evolution. Fossils now reveal that protoctists and the first animals—up to a meter across—existed millions of years before the so-called Cambrian explosion gave rise to trilobites, brachiopods, and other beloved creatures of our science textbooks. The earliest organisms of this Ediacaran period (type specimens hail to an unusually well-preserved formation in the Ediacara Hills of southern Australia) are oddly two dimensional, mouthless, and (most important) bereft of protective shells or carapaces. McMenamin and colleagues believe these creatures subsisted through symbiosis with photosynthetic or chemosynthetic bacteria embedded in their tissues. What put an end to this peaceful "Garden of Ediacara" (as McMenamin has named it)? Predation, pure and simple—and from that issued the adaptive value of mineralized coverings, which made brachiopods and trilobites so eminently preservable.

In this next essay, social scientist and polymath Peter A. Corning similarly cuts through the dualism of cooperation and competition. Interactions of both kinds can lead to synergism, relationships in which the

Peter A. Corning

One of the most ubiquitous aspects of the natural world is 'synergy'—combinatorial effects that are produced by the joint action of two or more discrete elements, components, or individuals. The term 'synergism' is a conceptual umbrella; like the term 'natural selection', synergism is not a mechanism or a controlling agency but a way of classifying an array of diverse effects in terms of a specific common property or characteristic. Synergism refers to co-operative effects of all kinds—effects that cannot be produced by the parts or individuals acting alone, or by any statistical summation of individual actions. Co-operation (hyphenated to emphasize the basic, functional connotation of the word) is not equivalent to altruism, and it is not a buzzword for 'good'. Although altruism and co-operation are often conflated, there can be both altruistic co-operation and egoistic co-operation (enlightened self-interest).

Synergistic effects are fundamental to molecular biology and biochemistry. Life itself involves synergistic combinations of, primarily, carbon, oxygen, nitrogen, and energy. Synergy is also evident in the basic building blocks of living systems. While DNA consists of only four alternative nucleotide bases, the precise patterning of the bases in various stable combinations of base pairs makes possible the co-operative construction of an endless variety of organic substances. Indeed, the functional significance of DNA emerges from its combinatorial properties. Likewise, DNA and RNA co-operate in constructing amino acids, with DNA serving as template and RNA as a vehicle for translation.

Synergism is commonplace at the cellular and organismic levels. One venerable example is teeth, which Aristotle invoked as evidence for design in nature; our teeth cannot perform independently of one another or of the supporting jaw structure and musculature. Other examples are the phenomena of summation in synaptic transmission, protein-metabolite interactions in cellular coordination activities, the coordinated construction and action of bone, muscle, and neuronal complexes, and the integrated functioning of the nervous system and major organ systems (visual, auditory, respiratory, digestive, neuroendocrine).

Synergism in Social Life

More germane to the present purpose is the growing recognition of synergism at the macro level—co-operative interactions between organisms of the same and different species which produce effects that could not otherwise be achieved. The functional significance of symbiosis is well established, and it is particularly important because it so clearly defies the "tooth and claw" model of natural selection. There are innumerable examples: the alga-fungus collaboration in lichen, where one partner specializes in photosynthesis and the other provides a secure attachment and water-retention capabilities that enable the two organisms together to occupy many otherwise uninhabitable areas; endoparasitic bacteria that aid the digestive processes of ruminants and termites by producing enzymes that break down cellulose; the blue-green algae that fix nitrogen for legumes and rice plants which in return provide an anoxic environment and/ or protective shade for the algae; the fungus-root symbioses of mycorrhizae; the Attini, several hundred ant species that maintain elaborate underground fungus "gardens"; ant-aphid symbioses, with the former specializing in defense and aphid nurturance and the latter in milking and processing the phloem sap of plants; the fifty-odd species of cleaner fish that earn their living by removing parasites from the bodies of larger fish; the bird species (such as cattle egrets and oxpeckers) that form attachments with large ruminants and specialize in removing ticks; and the sea anemone–crab partnerships, where the former provide camouflage and protection and the latter provide legs and mobility.

Perhaps the most striking examples are two colonial species that occupy a middle-ground position between symbionts and integrated organisms. One is the siphonophore *Physalia*, the Portuguese man-of-war, a floating colonial mass comprised of five different specialized polyps (a stolon plus stinging, feeding, reproductive, and protective tentacles). The other is the cellular slime mold *Dictyostelium*, which functions during part of its life cycle as a congeries of fully independent amoebae that forage for themselves and divide at frequent intervals. When food supplies diminish, the amoebae congregate into a sausage-shaped pseudoplasmodium that migrates like a multicellular organism and, in the final stage, transforms itself into an integrated fruiting body that produces spores.

Co-operative behavior is even more widespread among organisms of fully integrated species, and the range of functional interactions is diverse. They include: co-operative hunting and foraging, which may serve to increase capture efficiency, the size of the prey that can be pursued, or the likelihood of finding food patches; cooperative detection, avoidance of, and defense against predators, the forms of which range from mobbing and other kinds of coordinated attacks to flocking, communal nesting, and synchronized breeding; collective foraging and migration; co-operative competition, particularly in relation to obtaining food, territory, social dominance, and mates; co-operative protection of food supplies, notably among many insects and some birds that store food jointly; co-operative movement, including formations that increase aerodynamic or hydrodynamic efficiency and reduce individual energy costs and/or facilitate navigation; co-operative reproduction, which can include joint nest building, joint feeding, and joint protection of the young; co-operative thermoregulation, sometimes by sharing heat and sometimes by joint

Prairie dog town, with rattlesnake and a family of burrowing owls

cooling efforts; and co-operative environmental conditioning, such as joint detoxification of a hypotonic solution by flatworms.

Some species derive multiple advantages from co-operative behaviors. Honeybees, for instance, benefit from co-operation in foraging, defense, protection of food supplies, reproduction, migration, nest building, and thermoregulation, all within the context of specialized castes and an elaborate division of labor. Another example, in mammals, is molerats, but humankind is, of course, the most outstanding example. The psychologist Donald T. Campbell coined the term ultrasociality to describe such complexly organized species. Ultrasociality is a class that contains few members.

The precise degree of functional integration among co-operators can vary widely, from casual and loosely "facultative" relationships that are easily dissolved, at one extreme, to "obligative" relationships in which there is complete functional interdependency. There are also many intermediate forms in which breakdowns would be deleterious but not fatal. In the case of lichen symbiosis, for example, the constituent algae and fungi are capable of living independently and often do so, but together they are more efficient at performing complementary tasks and are therefore able to inhabit various otherwise inhospitable environments. At the other extreme, grass-eating cattle and wood-eating termites are utterly dependent on their parasites, without which they could not break down and digest cellulose. Between the two extremes are the oxpecker birds, which feed opportunistically on various prey, only some of which are the exoparasites of their ruminant hosts.

It is possible for synergistic relationships to arise voluntarily (nature's equivalent of democratic consent) or through coercion (as in the slave-making behaviors of various ant species.) However, the manner in which co-operative relationships are achieved does not necessarily map perfectly to the distribution of costs and benefits. The fungi that are carefully nurtured in underground gardens by the Attini may well be the net beneficiaries in terms of their absolute Darwinian fitness, just as the dairy cattle that have been enslaved and artificially bred by human beings have at least avoided extinction at the hands of humankind (the fate of a great many other large mammals). Are we using them, or are they using us?

Egoistic Co-operation

Social co-operation does not occur in a vacuum; it takes place against a background of direct or indirect competition. Competition may even be a major stimulus for co-operation. Ecological communities are not simply gladiatorial fields dominated by deadly competition; they are networks of complex interactions, of interdependent self-interests (as Darwin insisted) that require mutual adjustment and accommodation with respect to both the other cohabitants and the dynamics of the local ecosystem. The necessity for competition is one half of a duality, the other half of which includes many opportunities for mutually beneficial co-operation, including many tacit arrangements that amount to an ecological division of labor and specialization of systemic functions.

It is important to differentiate between altruism (self-sacrifice for another) and co-operation (operating together). Altruism describes a certain class of functional consequences to an individual or a group, and the concept is heavily value-laden. Co-operation describes a behavioral relationship in which there is coordination with respect to some goal-state, and the concept is devoid of intrinsic value. Often the twin dichotomies of egoism and altruism, on the one hand, and competition and co-operation, on the other, are treated as being equivalent when in fact they are not. Co-operative behaviors could be the result of altruism on the part of one animal toward another or toward a group; or they could be occasioned by the prospect of mutual advantage; or they could be motivated entirely by egoistic concerns.

In human societies such relationships are called markets, exchanges, co-operatives, corporations, contracts, and so on. I maintain that the engine of our cultural evolution has been synergistic effects based largely on egoistic co-operation, with altruistic behaviors playing a decidedly secondary role.

Edward O. Wilson launched his discipline-defining work *Sociobiology* (1975) with the startling assertion that altruism is "the central theoretical problem of sociobiology: how can altruism, which by definition reduces personal fitness, possibly evolve by natural selection?" The implication here is that social life is based primarily on altruism, and Wilson later adopts W. D. Hamilton's view that there are only three classes of social behavior: altruism (self-sacrifice for another), selfishness (raising one's own fitness at the expense of another), and spite (lowering one's own fitness in order to diminish that of another).

I submit that Hamilton, Wilson, and other early sociobiologists left out of their taxonomies (and their theoretical models of social behavior) the most important category of all, egoistic co-operation, in which one's own and others' needs are simultaneously or serially satisfied by joint, coordinated, or reciprocal actions. In fact this category is not entirely excluded by Wilson; it is merely disguised and hidden in the thickets of his massive tome. He discusses mutually beneficial forms of co-operation, including many explicit cases of synergistic behaviors, under at least four headings: (1) "adaptive advantages" of social life; (2) "mutualism" (positive symbiosis) between species; (3) "reciprocal altruism", Robert Trivers's well-known though misleading term for social behaviors that may seem to provide uncompensated benefits to unrelated individuals but in fact do not—reciprocal altruism is neither strictly reciprocal nor altruistic (it is a subset of the broader category of egoistic co-operation); and (4) "altruistic behaviors" that can also be given a mutualistic interpretation.

How can mutualistic, synergistic forms of social behavior be included in fact but excluded in theory? In large measure, the problem stems from the failure to differentiate between altruism and co-operation. The customary dichotomy of self-serving and group-serving behaviors leaves out behaviors that may involve both.

Synergism and Progressive Evolution

The concept of functional synergism is not new, and many theorists have noted that biological systems exhibit synergistic properties, but to my knowledge, none has recognized that synergistic effects are also causal, and of central importance. The key lies in recognizing that in evolution it is the functional effects or consequences of a phenomenon in terms of survival and reproduction that are central to what we call natural selection; proximate effects also serve as the ultimate causes of evolutionary changes. This is as true for synergistic effects as it is for any other functional effect.

Accordingly, the hypothesis is that it is the selective advantages arising from various synergistic effects that constitute the underlying cause of the apparently orthogenetic (or directional) aspect of evolutionary history, that is, the progressive emergence of complex, hierarchically organized systems. Synergistic effects have been the key to the emergence of all biological organization. The principle is as applicable to the evolution of the eukaryotic cell and multicellular organisms as it is to colonial organisms, endosymbionts, exosymbionts, and animal and human societies; in other words, to cellular biology, organismic biology, sociobiology, and evolutionary anthropology.

A compelling piece of evidence is presented by the eukaryotic cell, a product of one of the earliest biological 'revolutions' in the direction of more complex organization. While the emergence of the eukaryote, the first structurally complex, nucleated cell, involved an intricate series of steps, there is now strong support for the hypothesis that the endosymbiosis between previously independent prokaryotic bacteria, blue-green algae, and incipient or recently nucleated cells was decisive. The evidence suggests that the mitochondria, chloroplasts, and perhaps other functionally specialized organelles found within eukaryotic cells had independent origins. Whether the union that led to eukaryotes was mutualistic or arose as a master-slave relationship, the result was beneficial to all parties; our cells are efficiently organized, functionally differentiated systems.

More evidence for synergism is found in another of the major biological revolutions, the emergence of metazoan (multicelled) creatures from single-celled protists some six hundred million years ago. Biologists generally agree that multicellular organisms evolved separately in each of the three higher kingdoms of life—fungi, plants, and animals—and that fungi and plants arose by the amalgamation of protist colonies. Sponges too are viewed as little more than confederations of protists. Only in the animal kingdom is the hypothesis of a symbiotic union still in question.

Biologist Earl Hanson continues to defend the minority view that multicellular animals arose through internal differentiation. However, laboratory techniques that are now available for comparing DNA of different life forms may soon clarify (if not resolve) the issue. If it can be shown that the marine flatworms that Hanson believes to be unique products of evolution by differentiation are in fact regressive forms that were derived from ancestral amalgamators, then it is likely that all of the higher life forms are the products of symbiosis, of functional synergism.

The synergism hypothesis may also account for social organization of various kinds. There are, for example, the protective advantages to the individual animal of the well-organized prairie dog towns; the benefits of flocking behavior among starlings; the so-called Fraser Darling Effect in a number of bird species, where the social context enhances individual reproductive stimulation; coalition behavior in a number of social carnivores and primates; pack hunting in killer whales and wild dogs; and the awesome defensive maneuvers of many ground-dwelling primates. There are many more examples in the literature, and no doubt more will come to light as further field studies are completed.

As Stephen Jay Gould puts it, even human beings may be the descendants of protistan colonies. Thus, from the earliest one-celled eukaryotes to the most advanced nation-states, complex biological systems may have been built up through a dual process of aggregation and progressive functional differentiation, and each major breakthrough to a more complex form of organization may have been 'directed' by opportunities for realizing functional synergism.

Beyond the Binary

In recent decades the role of co-operation in evolution has generally been downplayed and underrated. Many earlier writings on the subject are now mentioned only in footnotes, and the theme has been relatively neglected among mainstream evolutionists. It is a strange state of affairs, for which no single factor seems to have been responsible. But one major cause was the semantic and conceptual confusion that clothed such terms as co-operation and synergy. Darwinists and anti-darwinists alike have tended to identify natural selection with competition and to treat co-operation (read 'altruism' or 'morality') as something opposed to it.

The result was a polarization of viewpoints. Following the example set by Kropotkin, many theorists have reasoned that the existence of mutual aid either falsified the role of natural selection in evolution or, in the case of human evolution, set people apart from or above the natural world. By means of co-operation, the argument says, human beings have transcended "nature, red in tooth and claw."

Eagle Claw and Bean Necklace (Georgia O'Keeffe)

This view of co-operation was reinforced by the Social Darwinists and by various liberal economists who for decades emphasized the role of competition in economic life, generally giving co-operation no more than a bit part. Marxist economists have happily accepted this one-sided interpretation and turned it against the liberals. The Marxist position is that competition and social conflict are characteristic only of capitalist societies and that the transformation to socialism will bring about a co-operative society. Both viewpoints misinterpret the nature of both capitalist and extant socialist societies.

There is no inherent conflict between natural selection and co-operation, or between egoism and co-operation. Nor does co-operation necessarily require the kinds of motivations and behaviors that many social theorists equate with the term. There is not even an absolute dichotomy between competition and co-operation. Many social interactions involve co-operation in one area or with one group and competition in another. Darwin himself observed that co-operation for the purpose of engaging in competition has been a significant factor in sociocultural evolution.

The bottom line, though, is that in nature co-operation is a means to an end, and the end is synergy.

Cooperative synergy and competitive strife are thus the yin and yang, the anima and animus of biological evolution. It is perhaps human nature that drives us to enthrone one or the other as primary. Like hoops within hoops, what appears as competition at one level may be driven by or result in cooperation at the next level up or down, and vice versa. Perhaps it is one's attention to which of the two prevails in the outermost hoop—be that focus cells, organisms, species, or biosphere; individual humans, nation-states, or a global culture—that gives the tilt to our worldviews.

————

Whosoever wants to have right without wrong,
Order without disorder,
Does not understand the principles
Of heaven and earth.

—*Chuang-Tzu*

The previous two chapters, while in an obvious way polar, are similar in one respect. Whether selective forces arise out of strife or synergy, both chapters depict evolutionary change as something that happens to lineages of organisms—not something that lineages accomplish through their own efforts. Chance and contingency offer up the tableau of possibilities from which aimless and blind external forces sculpt the wonders of the living. The richness of anatomy simply reflects the richness of the living and nonliving surroundings and the whims of environmental change.

But what about will? What about volition? The textbook caricature of a Lamarckian giraffe straining to reach leafy treetops and thereby ensuring longer-necked progeny has long been discredited. But does will or initiative play a role in evolution in perhaps a more subtle sense?

An essay by Jacob Bronowski lays the groundwork for this chapter. Bronowski deals not with will, but with an automatic process that gives the buildup of complexity the look of will.

Stratified Stability and Unbounded Plans

Jacob Bronowski

The theory of evolution by natural selection was the most important single scientific innovation in the nineteenth century. When all the foolish wind and wit that it raised had blown away, the living world was different because it was seen to be a world in movement.

Polymath Jacob Bronowski (1908–1974) is best known for his 1973 book and television series The Ascent of Man, *from which the above quotation is drawn. Playing on the title of Darwin's 1871* The Descent of Man, *Bronowski explores how something as godlike as the human intellect might have arisen from humble ancestors. His answer is a process he calls "stratified stability." To explain stratified stability, Bronowski invokes a metaphorical ratchet that prevents evolutionary gains in complexity from backsliding.*

There is a relation between the direction of evolution and the direction of time. In a history of three thousand million years, evolution has not run backward—at least, by and large, and in a definable statistical sense, it has not run backward. Why is this? Why does evolution not run at random hither and thither in time? What is the screw that moves it forward,

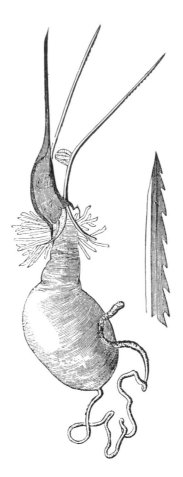

or a least, what is the ratchet that keeps it from slipping back? Is it possible to have such a mechanism which is not planned? What is the relation that ties evolution to the arrow of time, and makes it a barbed arrow?

Excerpts that follow are drawn from Jacob Bronowski's essay "New Concepts in the Evolution of Complexity: Stratified Stability and Unbounded Plans." This essay appeared in several publications noted in the bibliography. Permission for reprinting here is granted courtesy of Kluwer Academic Publishers, from the essay that appeared in volume 8 (1970) of Boston Studies in the Philosophy of Science.

Vitalism is a traditional and persistent belief that the laws of physics that hold in the inanimate world will not suffice to explain the phenomena of life. Of course it is not suggested, either by those who share the belief or by those like me who reject it, that we know all the laws of physics now, or will know them soon. Rather, what is silently supposed by both sides is that we know what kind of laws physics is made up of and will continue to discover in inanimate matter; and although that is a vague description to serve as a premise, it is what inspires vitalists to claim (and their opponents to deny) that some phenomena of life cannot be explained by laws of this kind.

The phenomena that are said to be inaccessible to physics are of two different kinds. One school of vitalists stresses the complexity of the individual organism. The other school of vitalists asserts that physical laws are insufficient to explain the direction of evolution in time: that is, the increase in complexity in new species, such as man.

Evolution has the direction, speaking roughly, from simple to more and more complex: more and more complex functions of higher organisms, mediated by more and more complex structures, which are themselves made of more and more complex molecules. How has this come about? How can it be explained if there is no overall plan to create more complex creatures—which means, at least, if there is no overall law (other than evolution as a mechanism) to generate complexity. In particular, how do we square this direction with the Second Law of Thermodynamics, which (as a general description subsuming ordinary physical laws) predicts the breakdown of complex structures into simple ones?

Evolution is commonly presented, even now, as if it required nothing but natural selection to explain its action, one minute step after another, as it were gene by gene. But an organism is an integrated system, and that implies that its coordination is easily disturbed. This is true of every gene: normal or mutant, it has to be incorporated into the ordered totality of the gene complex like a piece in a jigsaw puzzle. Yet the analogy of the jigsaw is too rigid: we need a geometrical model of stability in living processes (and in the structures that carry them out) which is not so landlocked against change. Moreover, the model must express the way in which the more complex forms of life arise from the simpler forms, and arise later in time. This is the model of stratified stability.

The Barb of Evolution

There are evolutionary processes in nature which do not demand the intervention of selective forces. Characteristic is the evolution of the chemical elements, which are built up in different stars step by step, first hydrogen to helium, then helium to carbon, and on to heavier elements. The encounter of hydrogen nuclei makes helium simply (though indirectly) because they hold together: arrangements are briefly formed which in time form the more complex configuration that is helium. Each helium nucleus is a new unit which is stable, and can therefore be used as a new raw material to build up still higher elements.

The most telling example is the creation of carbon from helium. Two helium nuclei which collide do not make a stable element, and fly apart

100μ.

100μ.

5μ.

Tapeworm in the intestine of a woodpecker

again in less than a millionth of a millionth of a second. But if in that splinter of time a third helium nucleus runs into the pair, it binds them together and makes a stable triad which is a nucleus of carbon. Every carbon atom in every organic molecule in every cell in every living creature has been formed by such a wildly improbable triple collision in a star.

Here then is a physical model which shows how simple units come together to make more complex configurations; how these configurations, if they are stable, serve as units to make higher configurations; and how these higher configurations again, provided they are stable, serve as units to build still more complex ones, and so on. Ultimately a heavy atom such as iron, and perhaps even a complex molecule containing iron (such as hemoglobin), simply fixes and expresses the potential of stability which lay hidden in the primitive building blocks of cosmic hydrogen.

The sequence of building up stratified stability is also clear in living forms. Atoms build the four base molecules, thymine and adenine, cytosine and guanine, which are very stable configurations. The bases are built into the nucleic acids, which are remarkably stable in their turn. And the genes are stable structures formed from the nucleic acids, and so on to the subunits of a protein, to the proteins themselves, to the enzymes, and step by step to the complete cell. The cell is so stable as a topological structure in space and time that it can live as a self-contained unit. Still, the cells in their turn build up the different organs which appear as stable structures in the higher organisms, arranged in different and more complex forms.

Two special conditions have assisted this mode of climbing from simple to complex. First, of course, there is the energy which comes to us from the sun, which increases the number of encounters between simple units and helps to lift them over the next energy barrier above them. (In the same way, simple atomic nuclei encounter one another reasonably often, and are lifted over the next energy barrier above them, by the energy in hot stars.) And second, natural selection speeds up the establishment of each new stratum of stability in the forms of life.

The stratification of stability is fundamental in living systems, and it explains why evolution has a consistent direction in time. Single mutations are errors at random, and have no fixed direction in time, as we know from experiments. And natural selection does not carry or impose a direction in time either. But the building up of stable configurations does have a direction, the more complex stratum built on the next lower, which cannot be reversed in general (though there can be particular lines of regression, such as the viruses and other parasites which exploit the more complex biological machinery of their hosts).

Here is the barb which evolution gives to time: it does not make it go forward, but it prevents it from running backward. The back mutations which occur cannot reverse it in general because they do not fit into the level of stability which the system has reached: even though they might offer an individual advantage to natural selection, they damage the organization of the system as a whole and make it unstable. Because stability is stratified, evolution is open, and necessarily creates more and more complex forms.

There is therefore a peculiar irony in the vitalist claim that the progress of evolution from simple to complex cannot be the work of chance. On the contrary, as we see, exactly this is how chance works and is constrained to work by its nature. The total potential of stability that is hidden in matter can only be evoked in steps, each higher layer resting on the layer below it. The stable units that compose one layer are the raw material for random encounters which will produce higher configurations, some of which will chance to be stable. So long as there remains a potential of stability which has not become actual, there is no other way for chance to go. It is as if nature were shuffling a sticky pack of cards, and it is not surprising that they hold together in longer and longer runs.

Thermodynamics and the Arrow of Time

It is often said that the progression from simple to complex runs counter to the normal statistics of chance that are formalized in the Second Law of Thermodynamics. Strictly speaking, we could avoid this criticism simply by insisting that the Second Law does not apply to living systems in the environment in which we find them. For the Second Law applies only when there is no overall flow of energy into or out of a system, whereas all living systems are sustained by a net inflow of energy.

But though this reply has a formal finality, in my view it evades the underlying question that is being asked. True, life could not have evolved in the absence of a steady stream of energy from the sun—a kind of energy wind on the earth. But if there were no more to the mechanism of molecular evolution than this, we should still be at a loss to understand how more and more complex molecules came to establish themselves. All that energy wind can do, in itself, is to increase the range and frequency of variation around the average state: that is, to stimulate the formation of more complex molecular arrangements. But most of these variant arrangements fall back to the norm almost at once, by the usual thermodynamic processes of degradation; so that it remains to be explained why they do not all do so, and how instead some complex arrangements establish themselves, and become the base for further complexity in their turn.

It is therefore relevant to discuss the Second Law, which is usually interpreted to mean that all constituent parts of a system must fall progressively to their simplest states. But this interpretation quite misunderstands the character of statistical laws in general in nonequilibrium states. The Second Law describes the final equilibrium state of a system; if we

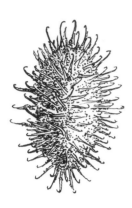

Seed-containing bur of *Xanthium spinosum*. This native of South America has hitched rides into North America, Europe, the Mediterranean region, and even Australia, where it is regarded as a weed.

are to apply it, as here, to stable states which are far from equilibrium, we must interpret and formulate it differently. In these conditions, the Second Law of Thermodynamics becomes a physical law only if there is added to it the condition that there are no preferred states or configurations.

In itself, the Second Law merely enumerates all the configurations which a system could take up, and it remarks that the largest number in this count are average or featureless. Therefore, if there are no preferred configurations (that is, no hidden stabilities in the system on the way to equilibrium), we must expect that any special feature that we find is exceptional and temporary, and will revert to the average in the long run. This is a true theorem in combinatorial arithmetic, and (like other statistical laws) a fair guess at the behavior of long runs. But it tells us little about the natural world which, in the years since the Second Law seemed exciting, has turned out to be full of preferred configurations and hidden stabilities, even at the most basic and inanimate level of atomic structure.

The Second Law describes the statistics of a system around equilibrium whose configurations are all equal, and it makes the obvious remark that chance can only make such a system fluctuate around its average. There are no stable states in such a system, and there is therefore no stratum that can establish itself; the system stays around its average only by a principle of indifference, because numerically the most configurations are bunched around the average.

But if there are hidden relations in the system on the way to equilibrium which cause some configurations to be stable, the statistics are changed. The preferred configurations may be unimaginably rare; nevertheless, they present another level around which the system can bunch, and there is now a countercurrent or tug-of-war within the system between this level and the average. Since the average has no inherent stability, the preferred stable configuration will capture members of the system often enough to change the distribution; and, in the end, the system will be established at this level as a new average. In this way, local systems of a fair size can climb up from one level of stability to the next, even though the configuration at the higher level is rare. When the higher level becomes the new average, the climb is repeated to the next higher level of stability; and so on up the ladder of strata.

When there are hidden strata of stability, one above another, as there are in our universe, it follows that the direction of time is given by the evolutionary process that climbs them one by one. Indeed, if this were not so, it would be impossible to conceive how the features that we remark could have arisen. We should have to posit a miraculous beginning to time at which the features (and we among them) were created ready-made, and left to fall apart ever since into a tohubohu of individual particles.

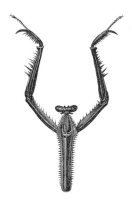

Head, thorax, and fore-
legs of a mantis

Time in the large, open time, takes its direction from the evolutionary processes which mark and scale it. So it is pointless to ask why evolution has a fixed direction in time, and to draw conclusions from the speculation. It is evolution, physical and biological, that gives time its direction; and no mystical explanation is required where there is nothing to explain. The progression from simple to complex, the building up of stratified stability, is the necessary character of evolution from which time takes its direction. And it is not a forward direction in the sense of a thrust toward the future, a headed arrow. What evolution does is to give the arrow of time a barb which stops it from running backward; and once it has this barb, the chance play of errors will take it forward of itself.

———

Having an environment, being in time,
And working out a history just
Like ourselves, God escapes
From the foreignness from all that is human,
Of the static timeless perfect absolute.

—*William James*

———

Self-Organization

As Jacob Bronowski pointed out, the second law of thermodynamics is widely (and wrongly) regarded as inimical to life. Misconstrued, the second law makes the origin of order, the rise of pulsing and purposeful cells, appear mysterious indeed. This outlook is changing, however. The final "heat death" of the universe still looms, but the inexorable press of entropy in the universe at large is now friend, not foe, of the scattered growth in complexity here and there. This new worldview, moreover, puts ordered entities in charge of their own evolution. No longer passive and pummeled into shape by a ruthless and fickle environment, living and quasi-living systems are seen as agents of their own fates.

"Self-organization" is an umbrella term that unites physicists, chemists, biologists, and even computer scientists who are searching for what underlies the growth of order and complexity in a universe otherwise governed by the second law's forces of dissolution. In the physical sciences it is not particularly controversial to study the self-organizing features of systems ranging from galaxies to whirlpools. The latter, for example, is a "dissipative structure" that builds and retains order by importing useful energy from the outside and exporting or dissipating energy in degraded form (entropy). But this kind of energy cream-skimming involves no will, no teleology: there is no ghost in the bathtub drain. Self-organization, in this sense, is a natural feature of the universe.

But self-organization is a controversial topic in biology. Fundamentally, many evolutionary biologists do not see any point in looking for ordering processes in living systems beyond what is already known. In their minds natural selection sifting through a rich offering of variation, and provided sufficient time, is perfectly capable of giving rise to everything awe-inspiring about life. It is natural selection, alone and unaided, that transformed organic molecules to cells, and cells to pluricellular organisms of all shapes and kinds.

Some biologists, however, are looking for more, for something perhaps deeper than natural selection, for an ordering process that happens within a living or quasi-living system before a new variation is served up to the grim reaper Darwin identified. The quest for new internal principles of ordering in biology is most intense at the border of the discipline: between cellular biology and organic chemistry. Specifically, origins-of-life researchers are finding tantalizing clues as to the kinds of processes that might have helped molecular building blocks over the hump of improbabilities that separates life from nonlife. Jacob Bronowski suggested that "stratified stability" would play a role, but other researchers have been making Bronowski's abstraction concrete. Notably, Manfred Eigen and colleagues have discovered how a "hypercycle" of molecules could arise and prosper as an interdependent and self-supporting group. If molecule

A catalyzes molecule B and B does C, and so on, somewhere down the line a molecule X might catalyze A. The chain then becomes a loop, and a gain made at any one step propagates throughout the cycle. The hypercycle thus shapes itself; it self-organizes.

The uroboros is a fitting metaphor for the paradigm of self-organization—and it is, in fact, the title of an international journal on the topic. The uroboros is an image of a serpent or dragon eating its tail. Borrowed from the Egyptians, it was used in medieval Europe as an alchemical symbol. In Carl Jung's interpretation, the dragon "devours, fertilizes, begets, slays, and brings itself to life again."

Biologist Stuart Kauffman, in his 1993 book The Origins of Order: Self-Organization and Selection in Evolution, *promotes a broadening of Darwinism to include self-organization:*

Stuart Kauffman

We have come to think of selection as essentially the only source of order in the biological world. If "only" is an overstatement, then surely it is accurate to state that selection is viewed as the overwhelming source of order in the biological world. It follows that, in our current view, organisms are largely ad hoc solutions to design problems cobbled together by selection. It follows that most properties which are widespread in organisms are widespread by virtue of common descent from a tinkered-together ancestor, with selective maintenance of the useful tinkerings. It follows that we see organisms as overwhelmingly contingent historical accidents, abetted by design.

It is not that Darwin is wrong, but that he got hold of only part of the truth. It is this single-force view which I believe to be inadequate, for it fails to notice, fails to stress, fails to incorporate the possibility that simple and complex systems exhibit order spontaneously. That spontaneous order exists, however, is hardly mysterious. The nonbiological world is replete with examples, and no one would doubt that similar sources of order are available to living things. What is mysterious is the extent of such spontaneous order in life and how such self-ordering may mingle with Darwin's mechanism of evolution—natural selection—to permit, or better, to produce what we see. Much of the order we see in organisms may be the direct result not of natural selection but of the natural order selection was privileged to act on.

Within the self-organization research community, Rod Swenson has made the most dramatic claim. While others search for material and efficient causes of self-organization (how this molecule catalyzes that one) and for formal causes (what kinds of interrelationships and feedbacks underlie the emergence of complexity), Swenson has been looking for the ultimate, the final cause. And he believes he has found it: this entropic universe is pocked by local regions of intense ordering (even life) because it is through ordered, dissipative systems that the rate of entropy production in the universe is maximized. The more life there is in the universe, the faster energy is degraded. The drive toward order, culminating perhaps in life, is thus a kind of shadow of the second law. Life does not flout the entropy principle; it serves it.

Overall, the self-organization perspective demands a fundamental shift in metaphysic. The aimless, aloof, and external power of natural selection (which itself overthrew the willful, paternal, but also external power of a transcendent god) is still important. But it is more the builder of particularity. The very framework of order in the universe and the evolutionary sequence of life owes to a kind of willful, self-sufficient, and (most important) internal power of self-organization.

Biochemist Bruce Weber and colleagues suggest that it is the residual hold of the Newtonian worldview that makes the self-organization perspective contrary to expectations. In a Newtonian universe objects and agents do not move or change course unless they are nudged from the outside. They do not move themselves. Inertia reigns. We thus demand the same kind of explanation of evolution as we demand of a thud in the dark. It is not natural, and we are not at peace until it is explained—and explained in comfortable Newtonian terms. But a self-organizing universe is one in which movement, ordering movement, is to be expected. In that kind of universe evolution is the lullaby of crickets outside the bedroom window.

The living thing is not the clay moulded by the potter,
Nor the harp played upon by the musician.
It is the clay modelling itself.

—Edward Stuart Russell

The Role of Initiative

Self-organization in the abstract, its role in assembling the building blocks of life, may seem far removed from the willfulness we humans know to be at the base of our own lives, our struggles to persist and grow, our quest for selfhood and community. Is there a role for something beyond automatic self-organization in the evolutionary process? Is there perhaps a role for initiative? Alister Hardy says yes.

Alister Hardy (1896–1985) was a professor of zoology at Oxford University. He made important technical contributions in marine ecology, but the bulk of his theoretical work and his popular writing dealt with evolutionary biology and its philosophical implications. In 1964 Hardy presented a series of invited lectures at the University of Aberdeen in Scotland. Known as the Gifford lectures for having been endowed by Lord Gifford, these events have given rise to several books that rank among the gems of Western thought and expression. Varieties of Religious Experience *by William James (1902) is among them. Lord Gifford specified in his will that the lectures should be aimed broadly at the topic of "natural theology." He further directed, "I wish the lecturers to treat their subject as a strictly natural science, the greatest of all possible sciences, indeed, in one sense, the only science, that of Infinite Being, without reference to or reliance upon any supposed special exceptional or so-called miraculous revelation. I wish it considered just as astronomy or chemistry is."*

Alister Hardy titled his compilation of Gifford lectures The Living Stream: Evolution and Man. *He explains, "We are part of a vast living stream flowing through time. At any moment now, if disaster does not overtake us, this living stream will hurl its first drops of humanity across the vault of space to land upon another planet. What power this living process has! What curiosity! What spirit!" Excerpts from* The Living Stream, *copyright 1965 by Sir Alister Hardy, are reprinted by permission of HarperCollins Publishers, Inc.*

Alister Hardy

I am a Darwinian in the modern sense, but I venture to suggest that there is something more about the process of evolution than is generally conceded by most biologists today; and that this "something more" does, I believe, link Natural Theology to the biological scheme. Out of this process of evolution, from somewhere has come the urge, or love of adventure, in Man that can drive him to risk his life in climbing Everest or in reaching the South Pole or the Moon. Is it altogether too naive to believe that this exploratory drive, this curiosity has had its beginnings in some deep-seated part of animal behaviour which is fundamental to the stream of life?

I am not a vitalist in the old-fashioned sense of the word. I fully expect that the whole of an animal's bodily mechanism will be resolved in terms of biophysics and biochemistry; but I am not a materialist in that I am blind to the reality of consciousness in the organic world. As yet we just do not know where, or how, it relates to the physiochemical system; and our science, at present, cannot deal with it. As a naturalist, however, I do not pretend that it does not exist, and I do not subscribe to the quite unproven dogma that consciousness only appeared at some point in the anthropoid stock. I would only remark that just as some physicists feel uncertain about the nature of gravity we would do well at present not to be dogmatic about the springs of animal behaviour.

Are natural selection and the gene complex the only factors [in evolution]? Important they must be—but are they all-important? Frankly we do not know, and I for one doubt it.

Plants, rooted to one spot, are certainly at the mercy of their environment; but animals which can move about have an opportunity to choose their habitat. In the evolution of motile animals there are two opposing selections at work. In the words of Mr. Charles Elton, there is "the selection of the environment by the animal as opposed to the natural selection of the animal by the environment."

Change of habit as a factor in evolution was of course the great contribution made to biological theory by Lamarck; a contribution which has been largely rejected or neglected. In stressing this factor Lamarck was, I believe, absolutely right. In saying this I must explain that I am not a Larmarckian in the generally accepted sense of the term; I do not believe that change of habit can influence evolution through a supposed inheritance of changes in bodily structure brought about directly by a greater use of some organs and a lesser use of others. I certainly feel, however, that Lamarck deserves much more credit, than he gets at present, for discerning the great importance of the behavioural side in the working of animal evolution. Again and again Lamarck made the point that changes in the environment can bring about changes in the habits of animals and that it is these changes of habit which can be so important in bringing about evolutionary modifications. I hope to show that there is good evidence that this is so, although not in the way that Lamarck thought; indeed that it is a cardinal part of the evolution process.

Behaviour as a Selective Force

How, in fact, do I believe that changes in behaviour alter the course of evolution? Quite simply by a form of Darwinian selection. My ideas are, I believe, a development of those first put forward by [J. Mark] Baldwin in America and [C.] Lloyd Morgan in this country at the end of the last century under the name of organic selection; some people, however, with whom I have discussed the matter insist that I am saying something rather different. At any rate I wish at the outset to acknowledge that my ideas spring from them and that I was directed to them by that wonderful mine of evolutionary knowledge and reference: Sir Julian Huxley's *Evolution, the Modern Synthesis*. The few biologists who have seriously considered Baldwin and Lloyd Morgan's organic selection as a factor in evolution have mostly done so in relation to habitat selection. I am concerned to show that it is a principle which may profoundly influence the evolution of the structure of animals.

If a population of animals should change their habits (no doubt often on account of changes in their surroundings such as food supply, breeding sites, etc., but also sometimes due to their exploratory curiosity discovering new ways of life, such as new sources of food or new methods of exploitation) then, sooner or later, variations in the gene complex will turn up in the population to produce small alterations in the animal's structure which will make them more efficient in relation to their new behaviour pattern; these more efficient individuals will tend to survive rather than the less efficient, and so the composition of the population

will gradually change. This evolutionary change is one caused initially by a change in behaviour.

If birds of a particular species, originally feeding on insects from the surface of the bark of trees, found, in a time of shortage, that they could get more prey by probing into or under the bark, then they might develop a change of habit which, by being copied by other members of the species, could gradually spread through the whole population. In recent years we have seen two examples of new habits spreading across the country: firstly the opening of milk bottles—first the cardboard tops, then the metal tops—spreading, apparently by copying right through the tit (*a kind of small bird*) populations of Europe, and secondly the spread across England of the attack on *Daphne* seeds by green-finches. When I mentioned my hypothetical birds pecking at insects on the bark of trees in a 1956 discussion at the Linnean Society, and then talked of tits opening milk bottles, someone half jokingly said "and eventually I suppose we shall have a race of tits with beaks like tin-openers!" Exactly. If milk bottles were some curious hard-covered organic objects and a species of

Blue tit and bottle, by
Margaret La Farge

tit specialised on them for food, then if bottles with thicker caps tended to survive, who could doubt that in time there would be evolved both thicker and thicker caps and more specialised tin-opener-like beaks for dealing with them?

Now to return to our birds probing into the bark for insects; if this new habit became well established and more profitable to the bird than mere pecking off the surface, then any members of the population with a gene complex giving a beak slightly better adapted to such probing would have a better chance of survival than those less well equipped. A new shape of beak would be evolved as a result of a change of habit. The same will apply to any other changes of habit as when an animal turns to digging for its food, diving into the water for fish and so on; in any population those gene complexes which modify particular organs to give a better expression to the new habit will, in the long run, supplant those which produced organs less efficient in satisfying needs.

Most people with whom I discuss this say that I am talking about pure Darwinism and seem to imply that I am just making a fuss about nothing. This is where I disagree and insist that there is a real if somewhat subtle difference: a difference which it is essential that we should understand if we are to appreciate the true nature of the living stream—the evolutionary process. To try and make clear this point I must return for a moment to the different kinds of selection.

We can classify the forms of selection in various ways. Most people when thinking of Darwinism think of two main types: selection by other organisms, including both predators and competitors, and selection by the inanimate environment. The first kind is usually subdivided into inter- and intra-specific selection, i.e. one of competition and combat between different species and the other of rivalry between members of the same species. Examples of the second kind of selection, that of the physical environment, are obvious, such as species of mammals with thicker fur being selected by survival in colder regions or races of wingless flies left surviving on small oceanic islands because those with normal wings are blown away to destruction.

Undoubtedly these kinds of selection account for a great deal of adaptation and evolutionary change. I would, however, make another kind of division between the forms of selection. I would distinguish all the foregoing kinds under a superheading of *external* selective agencies, meaning those acting from outside the organisms concerned, i.e. the selective forces acting from both the animate and inanimate environments; and in contrast to these I would place an *internal* selective force due to the behaviour and habits of the animal itself.

Much of behaviour—all instinctive behaviour—is, of course, governed by the gene complex. In addition, however, there are the kinds of changes in behaviour to which I have just referred: new modes of action which spread through the population and are maintained in higher vertebrates by tradition before they become converted by assimilation into instinctive action. I think it likely that most instinctive behaviour has developed in this way.

Now because a change of habit is usually occasioned by changes in the environment, it is generally supposed, I think, that any selection due to such a habit change is one differing only in degree, but not in kind, from the other forms of selection just discussed. This for me is the crux of the whole issue; I think they are radically different. I realise, of course, that it is the differential mortality in the population which brings about the survival of the more efficient type of beak, for example, and this is obviously mediated by factors in the external world killing off a higher proportion of the less efficient forms; nevertheless the real initiating agent in the process is the new behaviour pattern, the new habit.

I believe the case for regarding this "behavioural" type of selection as different in kind from the rest can be maintained. A new habit, as Lamarck said, is frequently the result of some environmental change and this may make this kind of selection seem similar to the other kinds. But among vertebrates, it must often be the restless, exploring and perceiving animal that discovers new ways of living, new sources of food, just as the tits have discovered the value of the milk bottles. The restless, exploratory type of behaviour has no doubt been fostered and developed by selection just because it pays dividends.

The Example of Darwin's Finches

Let me now continue with the subject of beaks. Darwin, writing in *The Voyage of the Beagle* of the finches of the Galapagos Islands, says:

The most curious fact is the perfect gradation in the size of the beaks in the different species of Geospiza, from one as large as that of a hawfinch to that of a chaffinch, and even to that of a warbler. The beak of Cactornis is somewhat like that of a starling; and that of the fourth sub-group, Camarhynchus, is slightly parrot-shaped.

Dr. David Lack, in his evolutionary study *Darwin's Finches* (1947) based upon his studies in the Galapagos, places the above quotation at the head of his chapter dealing with beak differences and food; I now give the following quotations from his study:

The chief way in which the various species of Darwin's finches differ from each other is in their beaks. Indeed, the beak differences are so pronounced that systematists have at various times used as many as seven different generic names for the birds. The beak differences between most of the genera and subgenera of Darwin's finches are clearly correlated with differences in feeding methods. This is well borne out by the heavy, finch-like beak of the seed-eating *Geospiza*, the long beak of the flower-probing *Cactornis*, the somewhat parrot-like beak of the leaf-, bud-, and fruit-eating *Platyspiza*, the woodpecker-like beak of the woodboring *Cac-*

tospiza, and the warbler-like beaks of the insect-eating *Certhidea* and *Pinaroloxias.*

Which is the more reasonable explanation of these adaptations; that chance mutations, first occurring in a few members of the population, caused these birds to alter their habits and seek new food supplies more suitable to their beaks and so become a more successful and surviving race, or did the birds, forced by competition, adopt new feeding habits which spread in the population so that chance changes in beak form giving greater efficiency came gradually to be preserved by organic selection? Differences of habit clearly play a great part in the ways of life of these different species. Can we really doubt which of the two explanations just suggested is the more likely to be true?

It is among these finches that we meet with one of the most remarkable examples of the exploring perceptive behaviour of animals. I quote again from Lack, from the same chapter.

The woodpecker-finch *Camarhynchus pallidus* has a stout, straight beak, with obvious affinities to that of the insectivorous tree-finches, but more elongated, and modified in the direction of that of a woodpecker or nuthatch. It feeds on beetles and similar insects, for which it searches bark and leaf clusters, and less commonly the ground, and also bores into wood. It is the only one of Darwin's finches to do this. It also possesses a remarkable, indeed a unique, habit. When a woodpecker has excavated in a branch for an insect, it inserts its long tongue into the crack to get the insect out. *C. Pallidus* lacks the long tongue, but achieves the same result in a different way. Having excavated, it picks up a cactus spine or twig, one or two inches long, and holding it lengthwise in its beak, pokes it up the crack, dropping the twig to seize the insect as it emerges.

The Role of Behaviour in Adaptive Radiation

Galapagos finches

I am not belittling the force of the external type of selection. We must always remember that both may be taking part together, but I would suggest that the behavioural influence, particularly in the higher groups, may often be the more fundamental in determining the animal's makeup; the effects of external selection are generally more limiting, pruning the organism to fit its surroundings or supplying it with the better means of escaping from its enemies. It is adaptations which are due to the animal's behaviour, to its restless exploration of its surroundings, to its initiative in seeking new sources of food when its normal supply fails or becomes scarce through competition, that distinguish the main diverging lines of evolution; it is these dynamic qualities which lead to the different roles of life that open up to a newly emerging group of animals in that phase of their expansion technically known as adaptive radiation.

Many authors have been both struck and puzzled at the rapidity with which the different orders of the placental mammals were evolved. Almost equally striking is the similar outburst exhibited by the reptiles in

Galapagos tortoises and
finches

the earlier Mesozoic age. G. G. Simpson discusses this problem in his *Tempo and Mode in Evolution* (1944), particularly the mammalian outburst, and explains it by supposing that at the time of such adaptive radiation the primitive mammals were split up into a large number of very small populations which would undergo more rapid evolution under the influence of "genetic drift". For such "drift" to operate, however, the populations would have to be exceedingly small. Normal selection, of course, would be more likely to produce many different lines if the mammalian stock were split up into small isolated populations, each being selected to suit the slightly different environments. I would suggest, however, that behavioural changes working through organic selection would give the more likely solution to the problem.

As the reptiles declined the small mammals multiplied and competed for the available food; with their exploratory behaviour they took to a great variety of new methods of obtaining a living: climbing, burrowing, swimming and flying. Mutation did not accelerate, but selection acted far more quickly as the changing habits developed; with keen competition in the new environments, behavioural selection quickly moulded the body forms of the different pioneering groups into the main new adaptive types. Once these were produced they remained characteristic of their particular modes of life and further modifications could only be of a minor kind.

It is a commonplace of evolutionary discussion to compare the similar adaptive radiation of the marsupial mammals isolated in Australia to that

Marsupial mammals of Australia and Tasmania. *Upper left:* Thylacine, or Tasmanian Tiger—probably extinct. *Bottom center:* the wolverinelike Tasmanian devil.

of the later and more typical placental mammals in the rest of the world; and particularly to point to the striking similarity in form of many of the corresponding types. The marsupials, cut off in Australia before the coming of the placentals, have developed nearly all the main terrestrial, adaptive types: herbivores and carnivores, arboreal and burrowing forms with one interesting exception: the leaping kangaroos and wallabies have developed in place of the swiftly running antelopes and deer elsewhere. Apart from the kangaroos, the similarity of the types of the marsupial and placental mammals is indeed remarkable.

What are the selective forces here? Are they simply those of similar environments or are they those of organic selection by similar behaviour patterns of animals obtaining the same kind of food and living in a similar fashion? I think it must be conceded that the latter is the more reasonable.

I think we can say, from the many different lines of argument, that the internal, behavioural selection, due to the "psychic life" of the animal, whatever we may think about its nature, is now seen to be a most powerful creative element in evolution.

———

I believe the first living cell
Had echoes of the future in it, and felt
Direction and the great animals, the deep green forest
And whale's-track sea; I believe this globed earth
Not all by chance and fortune brings forth her broods,
But feels and chooses. And the Galaxy, the firewheel
On which we are pinned, the whirlwind of stars in which our sun
Is one dust-grain, one electron, this giant atom of the universe
Is not blind force, but fulfills its life and intends its course.

—*Robinson Jeffers*

———

Changes in Preferences, Then Skills, Then Anatomy

A year after I came upon Alister Hardy's writings, I witnessed an extraordinary display of initiative and behavioral plasticity in the bird population along a remote stretch of the Gila River of New Mexico, where we rent a summer home. I watched an acorn woodpecker—who must have watched the hummingbirds—make reconnaissance flights to our feeder of sugar water. On about the fifth swoop she managed an awkward landing, clinging upside down, tail braced (though upward) in woodpecker fashion. Her long tongue was well adapted for this unusual food source, and her throat moved in rhythmic swallows for several minutes. A month earlier we had watched all sorts of birds become clumsy, though successful, flycatchers to take advantage of the mayfly hatch.

Finally, a pair of hepatic tanagers—the male a brilliant red, the female a stunning green and yellow—appeared one day in the scrub oak and juniper trees of our yard. I had seen none all summer, and in summers before I had caught only a few glimpses of these rare birds. The rains were late this year and the insects unusually scarce. This pair had obviously been watching the rufous and canyon towees make themselves conspicuous in order to have bread tossed out to them. The tanagers, too, sat in the usual spots, and when bread was tossed, these treetop insectivores (with two hungry juveniles soon to follow) became ground-feeding vegetarians.

In the remainder of this chapter Karl Popper, perhaps our century's best-known philosopher of science, explores the role of initiative in evolution. Popper agrees with Alister Hardy that behavioral initiative plays a profound role in evolutionary change. Both challenge the notion that passive organisms are simply pummeled by an active environment in the course of evolution. Hardy was, in fact, delighted to learn that Popper shared his viewpoint. Hardy recalls:

Alister Hardy

I used to enjoy, and still do, talking to undergraduate societies, but in discussion I would frequently be told, "Oh, I don't think Popper would agree with you there." I had read a lot of Sir Karl Popper's work and didn't really feel we were at odds with one another. Yet such remarks would continue to be made and I began to think of him as a kind of bogey-man I couldn't understand. The reader can imagine my delight when I received from Sir Karl a copy of his famous lecture, *Of Clouds and Clocks: an Approach to the Problem of Rationality and the Freedom of Man,* inscribed to me in these words: "With the expression of the greatest admiration for *The Living Stream*." Few intellectual events in my life have pleased me more.

Sir Karl's Of Clouds and Clocks *closes this chapter. But first, a short selection from his* Unended Quest: An Intellectual Autobiography *(copyright Karl R. Popper, 1976) presents the technical arguments underlying his view that behavioral initiative plays a significant role in evolutionary adaptation. It is reprinted here by permission of Open Court Publishing Company (La Salle, Illinois).*

Karl Popper

[Darwinism as a] theory predicts accidental mutations, and thus accidental changes. If any "direction" is indicated by the theory, it is that throw-back mutations will be comparatively frequent. Thus we should expect evolutionary sequences of the random-walk type. (A random walk is, for example, the track described by a man who at every step consults a roulette wheel to determine the direction of his next step.) Here an important question arises. How is it that random walks do not seem to be prominent in the evolutionary tree? The question would be answered if Darwinism could explain "orthogenetic trends," as they are sometimes called; that is, sequences of evolutionary changes in the same direction (nonrandom walks). Various thinkers such as Schrödinger and Wadding-ton, and especially Sir Alister Hardy, have tried to give a Darwinian ex-planation of orthogenetic trends, and I have also tried to do so. My suggestions for an enrichment of Darwinism which might explain ortho-genesis are briefly as follows:

I distinguish external or environmental selection pressure from internal selection pressure. Internal selection pressure comes from the organism itself and, I conjecture, ultimately from its preferences (or "aims") though these may of course change in response to external change. I assume that there are different classes of genes: those which mainly control the *anatomy*, which I will call *a*-genes; those which control *behaviour*, which I will call *b*-genes. The *b*-genes in their turn may be similarly subdivided into *p*-genes (controlling *preferences* or "aims") and *s*-genes (controlling *skills*).

Environmental changes may lead to new problems and so to the adop-tion of new preferences or aims (for example, because certain types of food have disappeared). The new preferences or aims may at first appear in the form of new tentative behaviour permitted but not fixed by the *b*-genes. In this way the animal may tentatively adjust itself to the new situation without genetic change. But this purely behavioural and tenta-tive change, if successful, will amount to the adoption, or discovery, of a new ecological niche. Thus it will favour individuals whose genetic *p*-structure (that is, their instinctive preferences or aims) more or less antici-pates or fixes the new behavioural pattern of preferences. This step will prove decisive; for now those changes in the skill structure (*s*-structure) will be favoured which conform to the new preferences: skills for getting the preferred food, for example.

I suggest that only after the *s*-structure has been changed will certain changes in the *a*-structure be favoured; that is, those changes in the ana-tomical structure which favour the new skills. The internal selection pressure in these cases will be "directed", and so lead to a kind of ortho genesis. My suggestion for this internal selection mechanism can be put schematically as follows: $p \rightarrow s \rightarrow a$. That is, the preference structure and its variations control the selection of the skill structure and its varia-tions; and this in turn controls the selection of the purely anatomical structure and its variations. This sequence, however, may be cyclical: the new anatomy may in its turn favour changes of preference, and so on.

What Darwin called "sexual selection" would, from the point of view expounded here, be a special case of the internal selection pressure which I have described; that is, of a cycle starting with new preferences. As an example of nonsexual selection I may mention the woodpecker. A reason-able assumption seems to be that this specialization started with a change in taste (preferences) for new foods which led to genetic behavioural changes, and then to new skills, and that the anatomical changes came last. A bird undergoing anatomical changes in its beak and tongue without undergoing changes in its taste and skill can be expected to be eliminated quickly by natural selection, but not the other way round.

The theory sketched suggests something like a solution to the problem of how evolution leads towards what may be called "higher" forms of life. Darwinism as usually presented fails to give such an explanation. It can at best explain something like an improvement in the degree of adaptation. But bacteria must be adapted at least as well as men. At any rate, they have existed longer, and there is reason to fear that they will survive men. But what may perhaps be identified with the higher forms of life is a behaviourally richer preference structure—one of greater scope; and if the preference structure should have (by and large) the leading role I ascribe to it, then evolution towards higher forms may become understandable.

The Critical Attitude and Cultural Evolution

Sir Karl Popper (born in 1902) is still active in the philosophy of science and evolutionary biology. His dictum of "falsifiability" rescued the scientific endeavor from the extremes of nineteenth-century logical positivism and the rampant subjectivity of our own era. In Popper's view, even the best scientific hypotheses and theories can never be proved true, absolutely and once-and-for-all. Rather, a scientific idea is distinguished by its ability, in principle, to be shown wrong. To be scientific, a new idea must be conducive to experimental test; it must be falsifiable. To be widely accepted, a scientific idea (a "conjecture" in Popper's words) must survive one or more serious attempts to refute it.

Conjectures and Refutations is, in fact, Popper's most widely cited book. In the excerpts that follow, you will see this theme (here termed "trial and error") carried through into Popper's rendition of Darwinian evolution, in both the biological and the cultural realm. Excerpts are drawn from his Compton Memorial Lecture, Of Clouds and Clocks: An Approach to the Problem of Rationality and the Freedom of Man *(copyright 1966 by Karl Raimund Popper), which was presented at and published by Washington University (St. Louis, Missouri).*

All organisms are constantly, day and night, engaged in problem-solving; and so are all those evolutionary sequences of organisms—the phyla which begin with the most primitive forms and of which the now living organisms are the latest members. Problem-solving always proceeds by the method of trial and error: new reactions, new forms, new organs, new modes of behavior, new hypotheses, are tentatively put forward and controlled by error elimination. Error-elimination may proceed either by the complete elimination of unsuccessful forms (the killing-off of unsuccessful forms by natural selection) or by the (tentative) evolution of controls which modify or suppress unsuccessful organs, or forms of behavior, or hypotheses.

The single organism is a kind of spearhead of the evolutionary sequence of organisms to which it belongs (its phylum): it is itself a tentative solution, probing into new environmental niches, choosing an environment and modifying it. It is thus related to its phylum almost exactly as the actions (behavior) of the individual organism are related to this organism: the individual organism and its behavior are both trials, which may be eliminated by error-elimination.

Evolution is clearly not a conscious process. Many biologists say that the evolution of certain organs solves certain problems; for example, that the evolution of the eye solves the problem of giving a moving animal a timely warning to change its direction before bumping into something hard. Nobody suggests that this kind of solution to this kind of problem is consciously sought. Is it not, then, just a metaphor if we speak of problem-solving?

I do not think so; rather, the situation is this: when we speak of a problem, we do so almost always from hindsight. Clearly it is in science that we are [in hindsight] most conscious of the problems we try to solve. So it should not be inappropriate to use hindsight in other cases, and to say that the amoeba solves some problems (though we need not assume that it is in any sense aware of its problems): from the amoeba to Einstein is just one step.

The amoeba's actions are not rational, while we may assume that Einstein's actions are. So there should be some difference, after all. I admit that there is a difference: even though their methods of trial and error movements are fundamentally not very different, there is a great difference in their attitudes towards error. Einstein, unlike the amoeba, consciously tried his best, whenever a new solution occurred to him, to fault it and detect an error in it: he approached his own solutions critically. I believe that this consciously critical attitude towards his own ideas is the one really important difference between the method of Einstein and that of the amoeba. It made it possible for Einstein to reject, quickly, hundreds of hypotheses as inadequate before examining one or another hypothesis more carefully, if it appeared to be able to stand up to more serious criticism.

As the physicist John E. Wheeler said recently, "Our whole problem is to make the mistakes as fast as possible." This problem of Wheeler's is solved by consciously adopting the critical attitude. This, I believe, is the highest form so far of the rational attitude, or of rationality. Since this attitude has led to the evolution of science, we may say that the argumentative function of language has created what is perhaps the most powerful tool for biological adaptation which has ever emerged in the course of organic evolution. Critical arguments are a means of control: they are a means of eliminating errors, a means of selection. We solve our problems by tentatively proposing various competing theories and hypotheses, as trial balloons, as it were; and by submitting them to critical discussion and to empirical tests, for the purpose of error-elimination.

The scientist's trials and errors consist of hypotheses. He formulates them in words, and often in writing. He can then try to find flaws in any one of these hypotheses, by criticizing it, and by testing it experimentally, helped by his fellow scientists who will be delighted if they can find a flaw in it. If the hypothesis does not stand up to these criticisms and to these tests at least as well as its competitors, it will be eliminated.

It is different with primitive man, and with the amoeba. Here there is no critical attitude, and so it happens more often than not that natural selection eliminates a mistaken hypothesis or expectation by eliminating those organisms which hold it, or believe in it. So we can say that the critical or rational method consists in letting our hypotheses die in our stead.

Like Descartes, I propose the adoption of a dualistic outlook, though I do not of course recommend talking of two kinds of interacting substances. But I think it is helpful and legitimate to distinguish two kinds of interacting states (or events), physio-chemical and mental ones. Moreover, I suggest that if we distinguish only these two kinds of states we still take too narrow a view of our world: at the very least we should also distinguish those artefacts which are products of organisms, and especially the products of our minds, and which can interact with our minds and thus with the state of our physical environment. Although these artefacts are often 'mere bits of matter', 'mere tools' perhaps, they are even on the animal level sometimes consummate works of art; and on the human level, the products of our minds are often very much more than 'bits of matter'—marked bits of paper, say; for these bits of paper may represent states of a discussion, states of the growth of knowledge, which may transcend (sometimes with serious consequences) the grasp of most or even all of the minds that helped to produce them. Thus we have to be not merely dualists, but pluralists; and we have to recognize that the great changes which we have brought about, often unconsciously, in our physical universe show that abstract rules and abstract ideas, some of which are perhaps only partially grasped by human minds, may move mountains.

Nature delights in progress; in advance
From worse to better: but when minds ascend,
Progress, in part, depends upon themselves.
Heaven aids exertion; greater makes the great;
The voluntary little lessens more.
O be a man! and thou shalt be a god!
And half self-made!—Ambition how divine!

—*Edward Young*

III *Embracing the Cosmos*

What disconcerts the modern world at its very roots is not being sure, and not seeing how it ever could be sure, that there is an outcome—a suitable outcome—to evolution. Half our present uneasiness would be turned to happiness if we could once make up our minds to accept the facts and place the essence and the measure of our modern cosmogonies within a noogenesis.

—*Pierre Teilhard de Chardin, 1955*

The ancient covenant is in pieces; man knows at last that he is alone in the universe's unfeeling immensity, out of which he emerged only by chance.

—*Jacques Monod, 1970*

The biospheric perspective admits human embeddedness in a global living system in a way that makes us neither so special as Teilhard portrays nor so arbitrary as Monod envisions. In the evolutionary long run, humanity itself will survive only as integral parts of a wild nexus of widely divergent life forms whose reproduction becomes possible only in concert.

—*Dorion Sagan, 1994*

Cosmic evolution, creative evolution, emergent evolution, noogenesis: these are the labels affixed to philosophies that bind biological evolution on earth with the rhythms of a progressive universe. In 1907 French philosopher Henri Bergson produced a book that launched "creative evolution" as a worldview. Fifteen years later biologist C. Lloyd Morgan used his prodigious understanding of science to give a progressive, cosmic view of evolution an expression more in accord with the leanings of fellow scientists. Morgan's vision of "emergent evolution" dispensed with Bergson's mysterious "élan vital"; complexity emerged through utterly material processes. "Noogenesis" was coined in 1925 by the French geologist and paleontologist Pierre Teilhard de Chardin. By 1940 Teilhard had produced a manuscript that gave full expression to his views, but it was only published posthumously, fifteen years later.

Bergson's book won a Nobel Prize for literature, and Morgan's was born out of a prestigious invitation to serve as a Gifford lecturer. But Teilhard's vision has had an effect that far surpasses the combined impact of all other books in the genre. Perhaps what makes Teilhard's thesis so compelling is his attention to the future. Other authors present the grand scheme of cosmic and biological evolution culminating in humankind, but Teilhard conjures an even greater grandeur that lies ahead. Perhaps, too, some readers are attracted to a philosophical extension of science that can mesh with Christianity. For Teilhard was not only a scientist; he was a Jesuit priest. Although "God" is not mentioned until the very end of his book, the epilogue is titled "The Christian Phenomenon." There Teilhard unifies his vision of a progressively evolving universe with a personal God and Christ.

Teilhard's synthesis of science and Christianity has appealed (indeed, brought epiphanies) to many, but it failed one big test: the Vatican forbade Teilhard to teach or publish his views. Upon his death, however, Teilhard's admirers were free to do what Teilhard wasn't. The book was published in France almost immediately (1955), and an English translation appeared four years later.

Le Phénomène Humain (translated in 1959 not as The Human Phenomenon *but as* The Phenomenon of Man*) is by no means easy to read. Although portions are inspired and poetic, big chunks can test the perseverance of even committed readers. Indeed, on my first attempt I could* *not get past the first several chapters in which Teilhard ploddingly (to my mind) lays out the physical and biological bases and the terminology for his worldview. Two years later, however, I skipped to the heart of the book and was amazed to find Teilhard more a delight than a struggle. Excerpts here are chosen for clarity, with an aim to convey the spirit as well as the substance of Teilhard's all-embracing vision.*

Julian Huxley wrote the introduction to the first English translation. There he recalls meeting Teilhard in 1946 and discovering that "he and I were on the same quest, and had been pursuing parallel roads ever since we were young men in our twenties." Huxley, a staunch proponent of evolutionary humanism and grandson of the first self-proclaimed "agnostic" (Thomas Huxley, who coined the term), did not share in Teilhard's theistic extensions. Huxley demurred, "Though many scientists, as I do, find it impossible to follow him all the way in his gallant attempt to reconcile the supernatural elements of Christianity with the facts and implications of evolution, this in no way detracts from the positive value of his naturalistic general approach." Overall, Huxley judges that Teilhard "has forced theologians to view their ideas in the new perspective of evolution, and scientists to see the spiritual implications of their knowledge."

Excerpts from The Phenomenon of Man *by Pierre Teilhard de Chardin (copyright 1955 by Editions du Seuil and copyright 1959 by William Collins Sons and Company, Ltd., London, and Harper and Row Publishers, New York) are reprinted by permission of HarperCollins Publishers, Inc. and Editions du Seuil.*

Body and Soul

Pierre Teilhard de Chardin

On the scientific plane, the quarrel between materialists and the upholders of a spiritual interpretation still endures. After a century of disputation each side remains in its original position and gives its adversaries solid reasons for remaining there.

So far as I understand the struggle, in which I have found myself involved, it seems to me that its prolongation depends less on the difficulty that the human mind finds in reconciling certain apparent contradictions in nature—such as mechanism and liberty, or death and immortality—as in the difficulty experienced by two schools of thought in finding a common ground. On the one hand the materialists insist on talking about objects as though they only consisted of external actions in transient relationships. On the other hand the upholders of a spiritual interpretation are obstinately determined not to go outside a kind of solitary introspection in which things are only looked upon as being shut in upon themselves in their 'immanent' workings. Both fight on different planes and do not meet; each only sees half the problem.

I am convinced that the two points of view require to be brought into union, and that they soon will unite in a kind of phenomenology or generalized physic in which the internal aspect of things as well as the external aspect of the world will be taken into account. Otherwise, so it seems to me, it is impossible to cover the totality of the cosmic phenomenon by one coherent explanation such as science must try to construct.

There is no concept more familiar to us than that of spiritual energy, yet there is none that is more opaque scientifically. On the one hand the objective reality of psychical effort and work is so well established that the whole of ethics rests on it and, on the other hand, the nature of this inner power is so intangible that the whole description of the universe in mechanical terms has had no need to take account of it, but has been successfully completed in deliberate disregard of its reality.

The difficulties we still encounter in trying to hold together spirit and matter in a reasonable perspective are nowhere more harshly revealed. Nowhere either is the need more urgent of building a bridge between the two banks of our existence—the physical and the moral—if we wish the material and spiritual sides of our activities to be mutually enlivened. To connect the two energies, of the body and the soul, in a coherent manner: science has provisionally decided to ignore the question, and it would be very convenient for us to do the same. Unfortunately, or fortunately, caught up as we are here in the logic of a system where the 'within' of things has just as much or even more value than their 'without', we collide with the difficulty head on. It is impossible to avoid the clash: we must advance.

What makes the crux—and an irritating one at that—of the problem of spiritual energy for our reason is the heightened sense that we bear without ceasing in ourselves that our action seems at once to depend on, and yet to be independent of, material forces. This is depressingly and magnificently obvious. 'To think, we must eat.' That blunt statement expresses a whole economy, and reveals, according to the way we look at it, either the tyranny of matter or its spiritual power. The loftiest speculation, the most burning love are, as we know only too well, accompanied and paid for by an expenditure of physical energy.

Once again: 'To think, we must eat.' But what a variety of thoughts we get out of one slice of bread! Like the letters of the alphabet, which can equally well be assembled into nonsense as into the most beautiful poem, the same calories seem as indifferent as they are necessary to the spiritual values they nourish.

To avoid a fundamental dualism, at once impossible and anti-scientific, and at the same time to safeguard the natural complexity of the stuff of the universe, I accordingly propose the following as a basis for all that is to emerge later. We shall assume that, essentially, all energy is psychic in nature; but add that in each particular element this fundamental energy is divided into two distinct components: a 'tangential energy' which links the element with all others of the same order (that is to say, of the same complexity and the same centricity) as itself in the universe; and a 'radial

*Reunion of the Soul
and the Body* (William
Blake)

energy' which draws it towards ever greater complexity and centricity—
in other words forwards.

A Line of Progress

Anyone who wishes to think in terms of evolution, or write about it,
should start off by wandering through one of those great museums—
there are four or five in the world—in which (at the cost of efforts whose
heroism and spiritual value will one day be understood) a host of travel-
lers has succeeded in concentrating in a handful of rooms the entire spec-

trum of life. There, without bothering about names, let him surrender himself to what he sees around him, and become impregnated by it: by the universe of the insects; by the molluscs, inexhaustibly variegated in their marblings and their convolutions; by the fishes, unexpected, capricious, as prettily marked as butterflies; by the birds, hardly less extravagant, of every form, feather, and beak; by the antelopes of every coat, carriage, and diadem. And so on, and so on. And for each word, which brings to our minds a dozen manageable forms, what multiplicity, what impetus, what effervescence! And to think that all we see are merely the survivors!

That there is an evolution of one sort or another is now common ground among scientists. Whether or not that evolution is directed is another question. Asked whether life is going anywhere at the end of its transformations, nine biologists out of ten will today say no, even passionately. They will say: "It is abundantly clear to every eye that organic matter is in a state of continual metamorphosis, and even that this metamorphosis brings it with time towards more and more improbable forms. But what scale can we find to assess the absolute or even relative value of these fragile constructions? By what right, for instance, can we say that a mammal, or even man, is more advanced, more perfect, than a bee or a rose? One spoke of the wheel is as good as any other; no one of the lines appears to lead anywhere in particular."

Contemplated without any guiding thread, it must be recognised that the host of living creatures forms qualitatively an inextricable labyrinth. What is happening, where are we going through this monotonous succession of ramifications? In the course of ages, doubtless, creatures acquire more organs of increased sensibility. But they also reduce them by specialisation. Besides, what is the real meaning of the term 'complication'? There are so many different ways in which an animal can become less simple—differentiation of limbs, of tissues, of sensory organs, of integument. All sorts of distributions are possible. Is there really one which can be said to be truer than the others? Is there one, that is to say, which gives to the whole of living things a more satisfying coherence, either in relation to itself, or in relation to the world to which life finds itself committed?

Leaving aside all anthropocentrism and anthropomorphism, I believe I can see a direction and a line of progress for life, a line and a direction which are in fact so well marked that I am convinced their reality will be universally admitted by the science of tomorrow.

Threshold of Reflection

Let us just try to see whether, amongst all the combinations tried out by life, some are not organically associated with a positive variation in the psychism of those beings which possess it. If so, let us seize on them and follow them; for, if my hypothesis be correct, they are undoubtedly the ones which, among the equivocal mass of insignificant transformations, represent the very essence of complexity, of essential metamorphosis. There is every chance that they will lead us somewhere.

Framed in these terms, the problem is immediately solved. Of course there exists in living organisms a selective mechanism for the play of consciousness. We have merely to look into ourselves to perceive it—the nervous system. We are in a positive way aware of one single interiority in the world: our own directly, and at the same time that of other men by immediate equivalence, thanks to language. But we have every reason to think that in animals too a certain inwardness exists, approximately proportional to the development of their brains. So let us attempt to classify living beings by their degree of 'cerebralisation'. What happens? An order appears—the very order we wanted—and automatically. Not only does the arrangement of animal forms according to their degree of cerebralisation correspond exactly to the classification of systematic biology, but it also confers on the tree of life a sharpness of feature, an impetus, which is incontestably the hall-mark of truth. Such coherence could not be the result of chance.

Among the infinite modalities in which the complication of life is dispersed, the differentiation of nervous tissue stands out, as theory would lead us to expect, as a significant transformation. It provides a direction; and therefore it proves that evolution has a direction. Since the natural history of living creatures amounts on the exterior to the gradual establishment of a vast nervous system, it therefore corresponds on the interior to the installation of a psychic state coextensive with the earth. On the surface, we find the nerve fibres and ganglions; deep down, consciousness.

If we wish to settle this question of the superiority of man over the animals (and it is every bit as necessary to settle it for the sake of ethics of life as well as for pure knowledge) I can only see one way of doing so—to brush resolutely aside all those secondary and equivocal manifestations of inner activity in human behaviour, making straight for the central phenomenon, reflection. From our experimental point of view, reflection is, as the word indicates, the power acquired by a consciousness to turn in upon itself, to take possession of itself as of an object endowed with its own particular consistence and value: no longer merely to know, but to know oneself; no longer merely to know, but to know that one knows. Admittedly, the animal knows. But it cannot know that it knows. In consequence it is denied access to a whole domain of reality in which we can move freely. We are separated by a chasm—or a threshold—which it cannot cross. Because we are reflective we are not only different but quite other. It is not merely a matter of change of degree, but of a change of nature, resulting from a change of state.

When water is heated to boiling point under normal pressure, and one goes on heating it, the first thing that follows—without change of temperature—is a tumultuous expansion of freed and vaporised molecules. Or, taking a series of sections from the base towards the summit of a cone, their area decreases constantly; then suddenly, with another infinitesimal displacement, the surface vanishes leaving us with a point. Thus by these remote comparisons we are able to imagine the mechanism involved in the critical threshold of reflection.

By the end of the Tertiary era, the psychical temperature in the cellular world had been rising for more than 500 million years. When the anthropoid had been brought mentally to boiling point some further calories were added. Or, when the anthropoid had almost reached the summit of the cone, a final effort took place along the axis. No more was needed for the whole inner equilibrium to be upset. What was previously only a centred surface became a centre. By a tiny 'tangential' increase, the 'radial' was turned back on itself and took an infinite leap forward. Outwardly, almost nothing in the organs had changed. But in depth, a great revolution had taken place: consciousness was capable of perceiving itself in the concentrated simplicity of its faculties.

Now the consequences of such a transformation are immense, visible as clearly in nature as any of the facts recorded by physics or astronomy. The being who is the object of his own reflection, in consequence of that very doubling back upon himself, becomes in a flash able to raise himself into a new sphere. In reality, another world is born. Abstraction, logic, reasoned choice and inventions, mathematics, art, calculation of space and time, anxieties and dreams of love—all these activities of inner life are nothing else than the effervescence of the newly-formed centre as it explodes onto itself.

Thenceforward it is easy to decide where to look in all the biosphere to see signs of what is to be expected. We already know that everywhere the active phyletic lines grow warm with consciousness towards the summit. But in one well-marked region at the heart of the mammals, where the most powerful brains ever made by nature are to be found, they become red hot. And right at the heart of that glow burns a point of incandescence. We must not lose sight of that line crimsoned by the dawn. After thousands of years rising below the horizon, a flame bursts forth at a strictly localised point. Thought is born.

A New Age

[There is] an unparalleled complexity of the human group—all those races, those nations, those states whose entanglements defy the resourcefulness of anatomists and ethnologists alike. There are so many rays in that spectrum that we despair of analysing them. Let us try instead to perceive what this multiplicity represents when viewed as a whole. If we do this we will see that its disturbing aggregation is nothing but a multitude of sequins all sending back to each other by reflection the same light. We find hundreds of thousands of facets, each expressing at a different angle a reality which seeks itself among a world of groping forms. We are not astonished (because it happens to us) to see in each person around us the spark of reflection developing year by year. We are all conscious, too, at all events vaguely, that something in our atmosphere is changing with the course of history. If we add these two pieces of evidence together, how is it that we are not more sensitive to the presence of something greater than ourselves moving forward within us and in our midst?

Under the free and ingenious effort of successive intelligences, something (even in the absence of any measurable variation of brain or cranium) irreversibly accumulates, according to all the evidence, and is transmitted, at least collectively by means of education, down the course of ages. The point here is that this something—construction of matter or construction of beauty, systems of thought or systems of action—ends up always by translating itself into an augmentation of consciousness, and consciousness in its turn, as we now know, is nothing less than the substance and heart of life in process of evolution. To this grand process it is fitting to apply with all its force the word 'hominisation'. Hominisation can be accepted in the first place as the individual and instantaneous leap from instinct to thought, but it is also, in a wider sense, the progressive phyletic spiritualisation in human civilisation of all the forces contained in the animal world.

How utterly warped is every classification of the living world in which man only figures logically as a genus or a new family. This is an error of perspective which deforms and uncrowns the whole phenomenon of the universe. To give man his true place in nature it is not enough to find one more pigeon-hole in the edifice of our systematisation or even an additional order or branch. With hominisation, in spite of the insignificance of the anatomical leap, we have the beginning of a new age. The earth 'gets a new skin'. Better still, it finds its soul.

There can indeed be no doubt that, to an imaginary geologist coming one day far in the future to inspect our fossilised globe, the most astounding of the revolutions undergone by the earth would be that which took place at the beginning of what has so rightly been called the psychozoic era. And even today, to a Martian capable of analysing sidereal radiations psychically no less than physically, the first characteristic of our planet would be, not the blue of the seas or the green of the forests, but the phosphorescence of thought.

Birth of the Noosphere

The biological change of state terminating in the awakening of thought does not represent merely a critical point that the individual or even the species must pass through. Vaster than that, it affects life in its organic totality, and consequently it marks a transformation affecting the state of the entire planet. Beneath the pulsations of geo-chemistry, of geo-tectonics and of geo-biology, we have detected one and the same fundamental process, always recognisable—the one which was given material form in the first cells and was continued in the construction of nervous systems. We saw geogenesis promoted to biogenesis, which turned out in the end to be nothing else than psychogenesis.

With and within the crisis of reflection, the next term in the series manifests itself. Psychogenesis has led to man. Now it effaces itself, relieved or absorbed by another and a higher function—the engendering and subsequent development of the mind, in one word 'noogenesis'.

Geologists have for long agreed in admitting the zonal composition of our planet. We have already spoken of the barysphere, central and metallic, surrounded by the rocky lithosphere that in turn is surrounded by the fluid layers of the hydrosphere and the atmosphere. Since [Edward] Suess, science has rightly become accustomed to add another to these four concentric layers, the living membrane composed of the fauna and flora of the globe, the biosphere, an envelope as definitely universal as the other 'spheres' and even more definitely individualised than them. For, instead of representing a more or less vague grouping, it forms a single piece of the very tissue of the genetic relations which delineate the tree of life.

A glow ripples outward from the first spark of self-reflection. The point of ignition grows larger. The fire spreads in ever widening circles till finally the whole planet is covered with incandescence. Only one interpretation, only one name can be found worthy of this grand phenomenon. Much more coherent and just as extensive as any preceding layer, it is really a new layer, the 'thinking layer', which, since its germination at the end of the Tertiary period, has spread over and above the world of plants and animals. In other words, outside and above the biosphere there is the noosphere.

To us, in our brief span of life, falls the honour and good fortune of coinciding with a critical change of the noosphere. In these confused and restless zones in which present blends with future in a world of upheaval, we stand face to face with all the grandeur, the unprecedented grandeur, of the phenomenon of man. What has made us in four or five generations so different from our forebears (in spite of all that may be said), so ambitious too, and so worried, is not merely that we have discovered and mastered other forces of nature. In final analysis it is, if I am not mistaken, that we have become conscious of the movement which is carrying us along, and have thereby realised the formidable problems set us by this reflective exercise of the human effort.

Obviously man could not see evolution all around him without feeling to some extent carried along by it himself. Darwin has demonstrated this. Nevertheless, looking at the progress of transformist views in the last hundred years, we are surprised to see how naively naturalists and physicists were able at the early stages to imagine themselves to be standing outside the universal stream they had just discovered.

Almost incurably, subject and object tend to become separated from each other in the act of knowing. We are continually inclined to isolate ourselves from the things and events which surround us, as though we were looking at them from outside, from the shelter of an observatory into which they were unable to enter, as though we were spectators, not elements, in what goes on. That is why, when it was raised by the concatenations of life, the question of man's origins was for so long restricted to the purely somatic and bodily side. A long animal heredity might well have formed our limbs, but our mind was always above the play of which it kept the score. However materialistic they might be, it did not occur to the first evolutionists that their scientific intelligence had anything to do in itself with evolution.

Man discovers that *he is nothing else than evolution become conscious of itself*, to borrow Julian Huxley's striking expression. It seems to me that our modern minds (because and inasmuch as they are modern) will never find rest until they settle down to this view. On this summit and on this summit alone are repose and illumination waiting for us.

Is evolution a theory, a system or a hypothesis? It is much more: it is a general condition to which all theories, all hypotheses, all systems must bow and which they must satisfy henceforward if they are to be thinkable and true. Evolution is a light illuminating all facts, a curve that all lines must follow.

The Source of Our Disquiet

It is impossible to accede to a fundamentally new environment without experiencing the inner terrors of a metamorphosis. The child is terrified when it opens its eyes for the first time. Similarly, for our mind to adjust itself to lines and horizons enlarged beyond measure, it must renounce the comfort of familiar narrowness. It must create a new equilibrium for everything that had formerly been so neatly arranged in its small inner world. It is dazzled when it emerges from its dark prison, awed to find itself suddenly at the top of a tower, and it suffers from giddiness and disorientation. The whole psychology of modern disquiet is linked with the sudden confrontation with space-time.

It cannot be denied that, in a primordial form, human anxiety is bound up with the very advent of reflection and is thus as old as man himself. Nor do I think that anyone can seriously doubt the fact that, under the influence of reflection undergoing socialisation, the men of today are particularly uneasy, more so than at any other moment of history. Conscious or not, anguish—a fundamental anguish of being—despite our smiles,

strikes in the depths of all our hearts and is the undertone of all our conversations. This does not mean that its cause is clearly recognised—far from it. Something threatens us, something is more than ever lacking, but without our being able to say exactly what. Let us try then, step by step, to localise the source of our disquiet.

In the first and most widespread degree, the 'malady of space-time' manifests itself as a rule by a feeling of futility, of being crushed by the enormities of the cosmos. Which of us has ever in his life really had the courage to look squarely at and try to 'live' a universe formed of galaxies whose distance apart runs into hundreds of thousands of light years? Which of us, having tried, has not emerged from the ordeal shaken in one or other of his beliefs? And who, even when trying to shut his eyes as best he can to what the astronomers implacably put before us, has not had a confused sensation of a gigantic shadow passing over the serenity of his joy?

Time and space are indeed terrifying. Accordingly what could make our initiation into the true dimensions of the world dangerous is for it to remain incomplete, deprived of its complement and necessary corrective—the perception of an evolution animating those dimensions. On the other hand, what matters the giddy plurality of the stars and their fantastic spread, if that immensity (symmetrical with the infinitesimal) has no other function but to equilibrate the intermediary layer where, and where only in the medium range of size, life can build itself up chemically? What matter the millions of years and milliards of beings that have gone before if those countless drops form a current that carries us along?

Our consciousness would evaporate, as though annihilated, in the limitless expansions of a static or endlessly moving universe. It is inwardly reinforced in a flux which, incredibly vast as it may be, is not only 'becoming' but 'genesis'. Indeed time and space become humanised as soon as a definite movement appears which gives them a physiognomy.

'There is nothing new under the sun' say the despairing. But what about you, O thinking man? Unless you repudiate reflection, you must admit that you have climbed a step higher than the animals. 'Very well, but at least nothing has changed and nothing is changing any longer since the beginning of history.' In that case, O man of the twentieth century, how does it happen that you are waking up to horizons and are susceptible to fears that your forefathers never knew?

What makes the world in which we live specifically modern is our discovery in it and around it of evolution. And I can now add that what disconcerts the modern world at its very roots is not being sure, and not seeing how it ever could be sure, that there is an outcome—a suitable outcome—to that evolution. In truth, half our present uneasiness would be turned into happiness if we could once make up our minds to accept the facts and place the essence and the measure of our modern cosmogonies within a noogenesis.

The Road Ahead

And now, by the very fact that we have measured the truly cosmic gravity of the sickness that disquiets us, we are put in possession of the remedy that can cure it. 'After the long series of transformations leading to man, has the world stopped? Or, if we are still moving, is it not merely in a circle?' Either nature is closed to our demands for futurity, in which case thought, the fruit of millions of years of effort, is stifled, still-born in a self-abortive and absurd universe. Or else an opening exists—that of the super-soul above our souls; but in that case the way out, if we are to agree to embark on it, must open out freely onto limitless psychic spaces in a universe to which we can unhesitatingly entrust ourselves.

Between these two alternatives of absolute optimism or absolute pessimism, there is no middle way because by its very nature progress is all or nothing. We are confronted accordingly with two directions and only two; one upwards and the other downwards, and there is no possibility of finding a half-way house. On neither side is there any tangible evidence to produce. Only, in support of hope, there are rational invitations to an act of faith.

Having got so far, what are the minimum requirements to be fulfilled before we can say that the road ahead of us is open? There is only one, but it is everything. It is that we should be assured the space and the chances to fulfill ourselves, that is to say, to progress till we arrive (directly or indirectly, individually or collectively) at the utmost limits of ourselves. This is an elementary request, a basic wage, so to speak, veiling nevertheless a stupendous demand. But is not the end and aim of thought that still unimaginable farthest limit of a convergent sequence, propagating itself without end and ever higher? Does not the end or confine of thought consist precisely in not having a confine? Unique in this respect among all the energies of the universe, consciousness is a dimension to which it is inconceivable and even contradictory to ascribe a ceiling.

There are innumerable critical points on the way, but a halt or a reversion is impossible, and for the simple reason that every increase of internal vision is essentially the germ of a further vision which includes all the others and carries still farther on. Having once known the taste of a universal and durable progress, we can never banish it from our minds any more than our intelligence can escape from the space-time perspective it once has glimpsed. If progress is a myth, that is to say, if faced by the work involved we can say: 'What's the good of it all?' our efforts will flag. With that the whole of evolution will come to a halt—because we are evolution.

Appearances of a Setback

At no previous period of history has mankind been so well equipped nor made such efforts to reduce its multitudes to order. We have 'mass movements'—no longer the hordes streaming down from the forests of the north or the steppes of Asia, but 'the Million' scientifically assembled. The Million in rank and file on the parade ground; the Million standardised in the factory; the Million motorised—and all this only ending up with Communism and National-Socialism and the most ghastly fetters. So we get the crystal instead of the cell; the ant-hill instead of brotherhood. Instead of the upsurge of consciousness which we expected, it is mechanisation that seems to emerge inevitably from totalisation.

In the presence of such a profound perversion of the rules of noogenesis, I hold that our reaction should be not one of despair but of a determination to re-examine ourselves. Monstrous as it is, is not modern totalitarianism really the distortion of something magnificent, and thus quite near to the truth? There can be no doubt of it: the great human machine is designed to work and must work—by producing a superabundance of mind. If it does not work, or rather if it produces only matter, this means that it has gone into reverse.

To explain or efface the appearances of a setback which, if it were true, would not only dispel a beautiful dream but encourage us to weigh up a radical absurdity of the universe, I would like to point out in the first place that to speak of experience—of the results of experience—in such a connection is premature to say the least. After all, half a million years, perhaps even a million, were required for life to pass from the pre-hominids to modern man. Should we now start wringing our hands because, less than two centuries after glimpsing a higher state, modern man is still at loggerheads with himself?

Each dimension has its proper rhythm. Planetary movement involves planetary majesty. Would not humanity seem to us altogether static if, behind its history, there were not the endless stretch of its prehistory? Similarly, and despite an almost explosive acceleration of noogenesis at our level, we cannot expect to see the earth transform itself under our eyes in the space of a generation. Let us keep calm and take heart.

The Personal and the Universal

When we listen to the disciples of Marx, we might think it was enough for mankind (for its growth and to justify the sacrifices imposed on us) to gather together the successive acquisitions we bequeath to it in dying—our ideas, our discoveries, our works of art, our example. Surely this imperishable treasure is the best part of our being.

Let us reflect a moment, and we shall soon see that for a universe which, by hypothesis, we admitted to be a collector and custodian of consciousness, the mere hoarding of these remains would be nothing but a colossal wastage. What passes from each of us into the mass of humanity by means of invention, education and diffusion of all sorts is admittedly of vital importance. I have sufficiently tried to stress its phyletic value and no one can accuse me of belittling it. But with that accepted, I am bound to admit that, in these contributions to the collectivity, far from transmitting the most precious, we are bequeathing, at the utmost, only the shadow of ourselves.

Our works? But even in the interest of life in general, what is the work of works for man if not to establish, in and by each one of us, an absolutely original centre in which the universe reflects itself in a unique and inimitable way? And those centres are our very selves and personalities. The exaltation, not merely the conservation, of elements by convergence: what, after all, could be more simple, and more thoroughly in keeping with all we know?

Is it not possible that in our theories and in our acts we have neglected to give due place to the person and the forces of personalisation? All our difficulties and repulsions as regards the opposition between the All and the Person would be dissipated if only we understood that, by structure, the noosphere (and more generally the world) represents a whole that is not only closed but also centred. Because it contains and engenders consciousness, space-time is necessarily of a convergent nature. Accordingly its enormous layers, followed in the right direction, must somewhere ahead become involuted to a point which we might call Omega, which fuses and consumes them integrally in itself.

Far from being mutually exclusive, the Universal and the Personal (that is to say, the 'centred') grow in the same direction and culminate simultaneously in each other. It is therefore a mistake to look for the extension of our being or of the noosphere in the Impersonal. The Future-Universal could not be anything else but the Hyper-Personal—at the Omega Point.

Thus it would be mistaken to represent Omega to ourselves simply as the centre born of the fusion of elements which it collects, or annihilating them in itself. By its structure Omega, in its ultimate principle, can only be a distinct Centre radiating at the core of a system of centres; a grouping in which personalisation of the All and personalisations of the elements reach their maximum, simultaneously and without merging, under the influence of a supremely autonomous focus of union.

Spiritual Renovation of Earth

First the molecules of carbon compounds with their thousands of atoms symmetrically grouped; next the cell which, within a very small volume, contains thousands of molecules linked in a complicated system; then the metazoa in which the cell is no more than an almost infinitesimal element; and later the manifold attempts made sporadically by the metazoa

to enter into symbiosis and raise themselves to a higher biological condition. And now, as a germination of planetary dimensions, comes the
thinking layer which over its full extent develops and intertwines in
fibres, not to confuse and neutralise them but to reinforce them in the
living unity of a single tissue.

I can see no coherent, and therefore scientific, way of grouping this
immense succession of facts but as a gigantic psycho-biological operation,
a sort of mega-synthesis, the super-arrangement to which all the thinking
elements of the earth find themselves today individually and collectively
subject. Mega-synthesis in the tangential, and therefore and thereby a
leap forward of the radial energies along the principal axis of evolution:
ever more complexity and thus ever more consciousness. If that is what
really happens, what more do we need to convince ourselves of the vital
error hidden in the depths of any doctrine of isolation? The egocentric
ideal of a future reserved for those who have managed to attain egoisti-
cally the extremity of 'everyone for himself' is false and against nature.
No element could move and grow except with and by all the others with
itself.

Also false and against nature is the racial ideal of one branch draining
off for itself alone all the sap of the tree and rising over the death of
other branches. To reach the sun nothing less is required than the combined growth of the entire foliage. The outcome of the world, the gates
of the future, the entry into the super-human—these are not thrown open
to a few of the privileged nor to one chosen people to the exclusion of all
others. They will open only to an advance of all together, in a direction
in which all together can join and find completion in a spiritual renovation of the earth.

Under conditions of distribution which in any other initial phylum would
have led long ago to the break-up into different species, the human verti-
cil as it spreads out remains entire, like a gigantic leaf whose veins, however distinct, remain always joined in a common tissue. With man we
find indefinite interfecundation on every level, the blending of genes, an-
astomoses of races in civilisations or political bodies. Zoologically speaking, mankind offers us the unique spectacle of a 'species' capable of
achieving something in which all previous species had failed. It has succeeded, not only in becoming cosmopolitan, but in stretching a single
organised membrane over the earth without breaking it.

The coalescence of elements and coalescence of stems, the spherical geometry of the earth and the psychical curvature of the mind harmonising
to counterbalance the individual and collective forces of dispersion in the
world and to impose unification—there at last we find the spring and
secret of hominisation. But with this point made, how are we to explain
the appearance all around us of mounting repulsion and hatred? If such a
strong potentiality is besieging us from within and urging us to union,
what is it waiting for to pass from potentiality to action? Just this, no

doubt: that we should overcome the 'anti-personalist' complex which paralyses us, and make up our minds to accept the possibility, indeed the reality, of some source of love and object of love at the summit of the world above our heads.

So long as it absorbs or appears to absorb the person, collectivity kills the love that is trying to come to birth. As such, collectivity is essentially unlovable. That is where philanthropic systems break down. Common sense is right. It is impossible to give oneself to an anonymous number. But if the universe ahead of us assumes a face and a heart, and so to speak personifies itself, then in the atmosphere created by this focus the elemental attraction will immediately blossom. Then, no doubt, under the heightened pressure of an infolding world, the formidable energies of attraction, still dormant between human molecules, will burst forth.

Love

There is, however, an obvious and essential proviso to be made. For the human particles to become really personalised under the creative influence of union—according to the preceding analysis—not every kind of union will do. Since it is a question of achieving a synthesis of centres, it is centre to centre that they must make contact and not otherwise. Thus, amongst the various forms of psychic inter-activity animating the noosphere, the energies we must identify, harness and develop before all others are those of an 'intercentric' nature, if we want to give effective help to the progress of evolution in ourselves. Which brings us to the problem of love.

Considered in its full biological reality, love—that is to say, the affinity of being with being—is not peculiar to man. It is a general property of all life and as such it embraces, in its varieties and degrees, all the forms successively adopted by organised matter. In the mammals, so close to ourselves, it is easily recognised in its different modalities: sexual passion, parental instinct, social solidarity, etc. Farther off, that is to say lower down on the tree of life, analogies are more obscure until they become so faint as to be imperceptible.

Driven by the forces of love, the fragments of the world seek each other so that the world may come to being. This is no metaphor; and it is much more than poetry. Whether as a force or a curvature, the universal gravity of bodies, so striking to us, is merely the reverse or shadow of that which really moves nature. To perceive cosmic energy 'at the fount' we must, if there is a 'within' of things, go down into the internal or radial zone of spiritual attractions. Love in all its subtleties is nothing more, and nothing less, than the more or less direct trace marked on the heart of the element by the psychical convergence of the universe upon itself.

Love alone is capable of uniting living beings in such a way as to complete and fulfil them, for it alone takes them and joins them by what is deepest in themselves. This is a fact of daily experience. At what moment do lovers come into the most complete possession of themselves if not when they say they are lost in each other? In truth, does not love every instant achieve all around us, in the couple or the team, the magic feat, the feat reputed to be contradictory, of personalising by totalising? And if that is what it can achieve daily on a small scale, why should it not repeat this one day on world-wide dimensions?

Mankind, the spirit of the earth, the synthesis of individuals and peoples, the paradoxical conciliation of the element with the whole, and of unity with multitude—all these are called Utopian and yet they are biologically necessary. And for them to be incarnated in the world all we may well need is to imagine our power of loving developing until it embraces the total of men and of the earth. It may be said that this is the precise point at which we are invoking the impossible. Man's capacity, it may seem, is confined to giving his affection to one human being or to very few. Beyond that radius the heart does not carry, and there is only room for cold justice and cold reason. To love all and everyone is a contradictory and false gesture which only leads in the end to loving no-one.

To that I would answer that if, as you claim, a universal love is impossible, how can we account for that irresistible instinct in our hearts which leads us towards unity whenever and in whatever direction our passions are stirred? A sense of the universe, a sense of the All, the nostalgia which seizes us when confronted by nature, beauty, music—these seem to be an expectation and awareness of a Great Presence. The mystics and their commentators apart, how has psychology been able so consistently to ignore this fundamental vibration whose ring can be heard by every practised ear at the basis, or rather at the summit, of every great emotion? Resonance to the All—the keynote of pure poetry and pure religion.

Summary

To make room for thought in the world, I have needed to 'interiorise' matter: to imagine an energetics of the mind; to conceive a noogenesis rising upstream against the flow of entropy; to provide evolution with a direction, a line of advance and critical points; and finally to make all things double back upon someone. In this arrangement of values I may have gone astray at many points. It is up to others to try to do better. My one hope is that I have made the reader feel both the reality, difficulty, and urgency of the problem and, at the same time, the scale and the form which the solution cannot escape. The only universe capable of containing the human person is an irreversibly 'personalising' universe. In such a vision man is seen not as a static centre of the world—as he for long believed himself to be—but as the axis and leading shoot of evolution, which is something much finer.

To the common sense of the 'man on the street' and even to a certain philosophy of the world to which nothing is possible save what has always been, perspectives such as these will seem highly improbable. But to a mind become familiar with the fantastic dimensions of the universe they will, on the contrary, seem quite natural, because they are simply proportionate with the astronomical immensities. In the direction of thought, could the universe terminate with anything less than the measureless—any more than it could in the direction of time and space?

Something will explode if we persist in trying to squeeze into our old tumble-down huts the material and spiritual forces that are henceforward on the scale of a world. A new domain of psychical expansion—that is what we lack. And it is staring us in the face if we would only raise our heads to look at it.

To outward appearance, the modern world was born of an anti-religious movement; man becoming self-sufficient and reason supplanting belief. Our generation and the two that preceded it have heard little but talk of the conflict between science and faith; indeed it seemed at one moment a foregone conclusion that the former was destined to take the place of the latter. But, as the tension is prolonged, the conflict visibly seems to need to be resolved in terms of an entirely different form of equilibrium—not in elimination, nor duality, but in synthesis.

After close on two centuries of passionate struggles, neither science nor faith has succeeded in discrediting its adversary. On the contrary, it becomes obvious that neither can develop normally without the other. And the reason is simple: the same life animates both. Neither in its impetus nor its achievements can science go to its limits without becoming tinged with mysticism and charged with faith. Like the meridians as they approach the poles, science, philosophy and religion are bound to converge as they draw nearer the whole.

———

The sun and stars that float in the open air;
The apple-shaped earth, and we upon it—surely the drift
   of them is something grand!
I do not know what it is, except that it is grand, and that it
   is happiness,
And that the enclosing purport of us here is not a speculation,
   or bon-mot, or reconnoisance,
And that it is not something which by luck may turn out well
   for us, and without luck must be a failure for us,
And not something which may yet be retracted in a certain
   contingency.

—*Walt Whitman*

Teilhard in the Ecological Age

Teilhard's vision may disappoint some of today's readers for its unmitigated anthropocentrism. But consider: Teilhard completed his manuscript in 1940, long before the crises of dwindling biodiversity, desertification, ozone depletion, global warming, and acid rain had implicated humankind as saboteurs, not saviors, of evolution. How might Teilhard have modified his views were he alive today? Thomas Berry, a cultural historian and president emeritus of the American Teilhard Association, makes just such an attempt in his 1982 pamphlet, Teilhard in the Ecological Age *(published by Anima Books). Berry insists that the Teilhardian vision is too important to be cast aside for its ecological insensitivity.*

Thomas Berry

The context which Teilhard provides is so comprehensive in its overarching perspectives and such an improvement on prior expressions of a Christian vision that we must press on in the basic context of his thought even while adapting his vision to a planetary crisis that he did not anticipate. The context provided by Teilhard is urgently needed to save the ecological movement from trivialization, while the ecological orientation is needed to save Teilhard from irrelevance in relation to the biological future of the earth community. To achieve these mutually beneficial results we must simply extend the principal concerns of Teilhard further than he himself extended them.

Thomas Berry proposes to "green" the Teilhardian vision in several ways. He suggests that the evolutionary process should be seen as finding "its highest expression in the earth community, not simply in a human community reigning in triumphal dominion." Thus "the human might better think of itself as a mode of being of the earth rather than simply as a separate being on the earth."

While Teilhard is indispensable in understanding the larger patterns of life orientation that are needed, his thought must not become fixated. A vision is needed that can provide the fullness of meaning and the energy required to activate a new human age in which peace with the earth will be established and the emergent creative process will move toward its next stage of fulfillment in the great age of the Earth Community that is before us.

A Secular Version of Cosmic Evolution

Julian Huxley joins Thomas Berry in giving an ecological gloss to Teilhard's cosmic vision. Huxley was, in fact, a leader of the conservation movement. He chaired a committee in 1947 whose work culminated in establishment of National Nature Reserves in Great Britain. About the same time, as first director-general of the United Nations Educational, Scientific, and Cultural Organization, he successfully pushed for creation of the International Union for the Conservation of Nature.

Julian Huxley

As a result of a thousand million years of evolution, the universe is becoming conscious of itself, able to understand something of its past history and its possible future. This cosmic self-awareness is being realized in one tiny fragment of the universe—in a few of us human beings. Perhaps it has been realized elsewhere too, through the evolution of conscious living creatures on the planets of other stars. But on this our planet, it has never happened before.

Evolution on this planet is a history of the realization of ever new possibilities by the stuff of which earth (and the rest of the universe) is made—life: strength, speed and awareness; the flight of birds and the social polities of bees and ants; the emergence of mind, long before man was ever dreamt of, with the production of colour, beauty, communication, maternal care, and the beginnings of intelligence and insight. And finally, during the last few ticks of the cosmic clock, something wholly new and revolutionary, human beings with their capacities for conceptual thought and language, for self-conscious awareness and purpose, for accumulating and pooling conscious experience. For do not let us forget that the human species is as radically different from any of the microscopic single-celled animals that lived a thousand million years ago as they were from a fragment of stone or metal.

The new understanding of the universe has come about through the new knowledge amassed in the last hundred years—by psychologists, biologists, and other scientists, by archaeologists, anthropologists, and historians. It has defined man's responsibility and destiny—to be an agent for the rest of the world in the job of realizing its inherent potentialities as fully as possible. It is as if man had been suddenly appointed managing director of the biggest business of all, the business of evolution—appointed without being asked if he wanted it, and without proper warning and preparation. What is more, he can't refuse the job. Whether he wants to or not, whether he is conscious of what he is doing or not, he *is* in point of fact determining the future direction of evolution on this earth. That is his inescapable destiny, and the sooner he realizes it and starts believing in it, the better for all concerned.

I believe that, by the time its implications have been properly grasped, the discovery of evolution is destined to have a more revolutionary effect upon ideology than any other scientific discovery yet achieved. The effect will also be a less depressing and more constructive one than that of astronomical discovery. For evolution bridges the gaps between man and animal, between mental and material, and between the organic and the inorganic. Evolution shatters the pretence of human isolationism and sets man squarely in his relation—and a very important relation—with the cosmos. It is the most powerfully integrative of concepts, forcibly and inevitably uniting nebulae and human emotions, life and its environment, religion and material nature, all into a single whole. The facts of evolution, once clearly perceived, indicate the position we men should take up and the function we are called on to perform in the universe. "Stand there," they say, "and do thus and thus." If we neglect to do as they order, we not only do so at our peril but are guilty of a dereliction of our cosmic duty.

The Sequence of Evolution

The process of evolution as we know it today exists on three distinct levels—the inorganic or cosmic, the biological or organic, and the human or psychosocial. There is complete continuity between the three phases or levels, but yet a critical point between each one and the next, after which the process alters in character.

Cosmic evolution means the process of change in the stars and nebulae, in the inorganic constituents of the cosmos, in so far as they are not caught up in the effects of the other phases of evolution. These constitute the enormous bulk of the whole; but their changes are slow almost beyond imagination, and the complexity of organization arrived at is almost infinitely below that produced by organic evolution in any familiar animal. Here and there in this dragging and apparently meaningless drama of lifeless matter, spots appear in which more complicated organizations of matter become possible, sometimes indeed organizations capable of self-copying and therefore alive. The astronomers tell us that we may expect that at least a few hundreds of these theatres of life have arisen in our own galaxy: but we have knowledge only about one—our own earth.

Living matter does not always copy itself exactly. From the basic fact of self-copying and the secondary fact of inaccurate copying or mutation, natural selection automatically follows; and natural selection is a far more rapid agency of change than anything available on the inorganic level. As a result of its operation during two thousand million years or so, it has in fact produced the organizations we call higher animals, which would, if they were not so familiar, stagger us by the almost impossible complexity and delicacy of their construction. A dog would be a miracle if it were not just our familiar Prince or Toby.

More extraordinary still, it has produced mental as well as physical organization. From the study of bacteria or amoebae, jellyfish or green plants, we would have no right to conclude that they possessed mental attributes: but with an ape or a cat or a bird we cannot avoid this conclusion. Their minds are certainly very different from ours, but minds they certainly possess.

Eventually the stream of evolutionary progress passed a second critical point, and a new level of evolution was attained. In one out of the million or so animal species, mind developed to a stage at which it gave its possessor the power for true speech and conceptual thought. The result was man. With this, a new method of evolutionary change was introduced—cumulative change in the behaviour and achievements of a social group by mentally transmitted tradition, instead of change in the potentialities of individuals by physically transmitted systems of nucleoproteins. And this again immensely speeded up the rate of the evolutionary process.

Evolution on the human level, although it has been operating for the barest fraction of geological time, has already produced very extraordinary new results, impossible even to conceive of on the biological level—for example, Dante's *Divina Commedia,* guided missiles, Picasso's *Guernica,* Einstein's theory of relativity, ritual cannibalism, the Parthenon, the Roman Catholic Church, the films of the Marx Brothers, modern textile mills, Belsen, and the mystical experiences of Buddhist saints. Most extraordinary in principle, it has generated values. No one can prove that values play a part in the process of biological evolution, but no one can deny that they do so in human affairs. In lower organisms, the only ultimate criterion is survival: but in man some experiences and actions, some objects and ideas, are valued for their own sake.

The ideologically most important fact about evolution is that the human species is now the spearhead of the evolutionary process on earth, the only portion of the stuff of which our planet is made which is capable of further progress. Men are the sole trustees, agents, representatives, embodiments, or instruments—each word has its merits and demerits—of the only process of progressive evolution with which we have any direct concern. It is thus part of human destiny to be the necessary agent of the cosmos in understanding more of itself, in bearing witness to its wonder, beauty, and interest, in creating new aids to and mechanisms for existence, in experiencing itself, and so introducing the cosmos to more new and more valuable experiences.

The Need for New Beliefs

Twentieth-century man, it is clear, needs a new organ for dealing with destiny, a new system of religious beliefs and attitudes adapted to the

new situation in which his societies now have to exist, including the new knowledge which they have discovered and amassed. The radically new feature of the present situation may perhaps be stated thus: Earlier religions and belief-systems were largely adaptations to cope with man's ignorance and fears, with the result that they came to concern themselves primarily with stability of attitude. But the need to-day is for a belief-system adapted to cope with his knowledge and his creative possibilities; and this implies the capacity to meet and to inspire change. In other words, the primary function of earlier systems was of necessity to maintain social and spiritual morale in face of the unknown: and this they accomplished with a considerable measure of success. But the primary function of any system today must be to utilize all available knowledge in giving guidance and encouragement for the continuing adventure of human development.

There should no longer be any talk of conflict between science and religion. Between scientific knowledge and certain religious systems, yes: but between science as an increasing knowledge of nature and religion as a social organ concerned with destiny, no. On the contrary, religion must now ally itself wholeheartedly with science. This is a task for the human species as a whole, to which all can bring their contribution. The co-operation of the religiously minded and the scientifically trained is essential for its adequate performance.

Many people will say that any deliberate attempt to create a belief-system is unnatural and doomed to failure; such things cannot be turned out artificially but must have a natural growth. I do not think that this is necessarily true. After all, the present epoch differs from all previous periods in possessing a far more extensive and detailed knowledge about the universe in general and about human societies in particular—though this knowledge must be synthesized and processed before it can serve as basis for a new belief-system. For this, new techniques of teamwork and group research and new forms of co-operation between specialisms will be required, as well as new types of educational curriculum and new techniques of teaching. In so far as an effective new belief-system must have a religious aspect, it will doubtless need to wait for the appearance of a prophet who can cast it into compelling form and shake the world with it.

The religion indicated by our new view of our position in the cosmos must be one centred on the idea of fulfilment. Man's most sacred duty, and at the same time his most glorious opportunity, is to promote the maximum fulfilment of the evolutionary process on this earth; and this includes the fullest realization of his own inherent possibilities.

An Evolutionary Ethic

There is inevitably some conflict between the interests of individuals and those of society. But the conflict is in large measure transcended in this conception of man as an evolving psycho-social organism. In the longest-term view, our aim must be to develop a type of society and culture capable of ever-fresh evolution, one which continually opens the way to new and fuller realizations; in the medium-term point of view, we must secure the reproduction and improvement of psycho-social organization, the maintenance of the frameworks of society and culture and their transmission and adjustment in time; and in the immediate point of view we must aim at maximum individual fulfillment.

What needs stressing, however, is that, from the angle of evolutionary humanism, the flowering of the individual is seen as having intrinsic value, as being an end in itself. In the satisfying exercise of our faculties, in the pure enjoyment of our experience, the cosmic process of evolution is bringing some of its possibilities to fruition. In individual acts of comprehension or love, in the enjoyment of beauty, in inner experiences of peace and assurance, in the satisfactions of creative achievement, however humble, we are helping to realize human destiny.

Above all, the individual should aim at fulness and wholeness of development. Every human being is confronted with the task of growing up, of building a personality out of the raw materials of his infant self. A rich and full personality, in moral and spiritual harmony with itself and with its destiny, one whose talents are not buried in a napkin, and whose wholeness transcends its conflicts, is the highest creation of which we have knowledge, and in its attainment the individual possibilities of the evolutionary process are brought to supreme fruition. But if the individual has duties towards his own potentialities, he owes them also to those of others, singly and collectively. He has the duty to aid other individuals towards fuller development, and to contribute his mite to the maintenance and improvement of the continuing social process, and so to the march of evolution as a whole.

Three main contrasted ideals of personal development are possible. One is specialization: the fullest exploitation of some particular capacity, as seen in many successful professional men. The second we may call all-roundedness by summation: the cultivation of every kind of fulfilment separately. This was, broadly, the ideal of the ancient Athenians and of our own Elizabethans. The third is difficult to characterize in a word: we may perhaps call it comprehensive wholeness: the cultivation of inner harmony and peace, the development of a unitary and comprehensive pattern of intellectual and spiritual organization. This has been the aim of the saints, the sages, and the mystics.

Restoring Our Unity with Nature

Evolutionary humanism, with its naturalism and its twin concepts of present fulfilment for the individual and of long-term progressive realization of possibilities for man and the planet he inhabits, imposes the need for a transvaluation of values. For one thing, it helps to restore our unity with nature. It brings back the objects of our adoration and the goals of our spiritual longings out of supernatural remoteness and sites them nearer home, in the immediacy of experience.

As an example, let us consider the beauty and richness of nature. Rare Christian mystics like Traherne have found in it a religious fulfilment, and great poets like Wordsworth succeeded in expressing its transcendent value. The gospel of evolutionary humanism generalizes that value. The enjoyment of the beauty and strange variety of the natural world—an experience engendered by nature and the capacities of man's mind—is seen as one of the indispensable modes of human fulfilment, not to be neglected without peril, involving something essentially religious or sacred even though we may not burden it with any such heavy designation. As a corollary, we have the collective duty of preserving nature—partly for its own sake, but mainly as one of the necessary means for man's fulfilment. To exterminate a living species, be it lion or lammergeier, to desecrate the landscape, to wipe out wild flowers or birds over great tracts of country, is to diminish the wonder, the interest, and the beauty of the universe.

Lammergeier

Thus, for man to fulfil his destiny, he must think of himself as in partnership with the cosmos. Just as he cannot exist adequately if he exhausts or overspends the material resources of the earth, so he cannot realize many possibilities of beauty or wonder if he too much destroys or tames the beauty, strangeness, and variety of nature—as by putting dams and pylons and bungalows all over Snowdonia or the sea-coast, or killing off big game, or draining every drainable pond or lake. Evolution thus insists on the oneness of man with nature, not merely in respect of biological descent and chemical composition, but because nature is the indispensable basis of his material existence, and also the indispensable partner in his mental and spiritual achievements.

The Scientific Spirit

Man finds one of his ultimate fulfilments in comprehension. Fuller comprehension is one of the basic duties (and privileges) of the individual. Secondly, accumulated and organized knowledge and experience are necessary instruments or organs of human advance. Thus scientific research in all fields is essential, and its encouragement is one of the most important tasks of a civilized society.

Science has also contributed a discovery of the first magnitude—the discovery of the principle of limited but increasing certitude as the best method of extending and organizing knowledge. The principle of limited certitude not only includes scientific method in the restricted sense—the method of dispassionate observation and, where possible, experiment, of framing hypotheses, and of their testing and modification in the light of further observation and experiment. But it comprises more than this: it involves a general attitude to experience. It implies a fundamental humility, in acknowledging at the outset our enormous ignorance, the vast extent of what we do not know. But it also implies a legitimate pride and assurance—pride in the extent of the areas already annexed to the domain of knowledge from the wastes of ignorance, assurance in the tested validity of the accumulated facts and in the efficacy of the scientific method.

The scientific spirit and the scientific method have proved the most effective agents for the comprehension and control of physical nature. It remains for man to apply them to the comprehension and control of human destiny. For this to happen, science must understand that religion of some kind is a necessary organ for coping with the problems of destiny; and religion must not only accept and utilize the findings of science, but must be willing to admit the central principle of limited certitude, with its implication of progressive but always incomplete achievement of a better religious construction.

On Population, Eugenics, and Art

There are a few other points on which I would like to touch. The importance of the population problem for human destiny is now beginning to loom large. The implications of evolutionary humanism here are clear. If the full development of human individuals and the fulfilment of human possibilities is the overriding aim of our evolution, then any over-population which brings malnutrition and misery, or which erodes the world's material resources or its resources of beauty or intellectual satisfaction, is evil. Among the world's major and immediate tasks is the working out of an effective and acceptable population policy. In the ultimate light of humanist values, the deliberate encouragement of over-population for military or political ends, as in pre-war Italy and Japan, the intellectual dishonesty of the Russian Communists in asserting that over-population is an invention of the "Morganist-Weismannist hirelings of American monopolists" designed to justify American imperialist expansion, and the theological dogmatism of the Roman Catholics which denounces birth-control and prevents the scientific discussion of population problems even in international bodies like the World Health Organization—all are seen as immoral and indeed wicked.

Evolutionary humanism has eugenic implications also. These are, for the moment, largely theoretical, but in due time will become immensely practical. Within a century we should have amassed adequate knowledge of what could be done negatively to lighten the burden of inherited deficiency of mind or body which presses so cruelly on so many individual human beings and so heavily on evolving humanity as a whole, and positively to raise the entire level of innate human possibilities and capacities.

When this has happened, the working out of an effective and acceptable eugenic policy will be seen as not only an urgent but an inspiring task, and its political or theological obstruction as immoral.

I must say a word about the arts. Art, science, and religion are the three main fields of man's creative activity: all are indispensable for his fulfilment and the greater realization of his possibilities. In its recent manifestations, Western civilization has tended to exalt science and its technological applications at the expense of the arts. But we can grasp how important and indispensable they really are by imagining a world without them. Think of a world without music or poetry, without its churches and noble houses, without ballets, plays, novels, and films, without pictures and sculptures: such a world would be intolerable and life in it unlivable.

Personal Reflections

In exposing my thesis, I have had to range discursively into many fields. In concluding, perhaps I may be permitted to bring them together in a personal focus. I can, at any rate, testify to the fact that the concept of evolutionary humanism has been of value to myself. It has enabled me to resolve many of the dilemmas and conflicts with which any enquiring and aspiring mind is inevitably beset. It has enabled me to see this strange universe into which we are born as a proper object both of awe and wondering love and of intellectual curiosity. More, it has made me realize that both my wonder and my curiosity can be of significance and value in that universe.

It has enabled me to relate my experiences of the world's delights and satisfactions, and those of its horrors and its miseries, to the idea of fulfilment, positive or negative. In the concept of increased realization of possibilities, it provides a common measuring rod for all kinds of directional processes, from the development of personal ethics to large-scale evolution, and gives solid ground for maintaining an affirmative attitude and faith, as against that insidious enemy, Goethe's *Geist, der stets verneint,* the spirit of negation and despair. It affirms the positive significance of effort and creative activity and enjoyment. In some ways most important of all, it has brought back intellectual speculation and spiritual aspiration out of the abstract and isolated spheres they once seemed to me to inhabit, to a meaningful place in concrete reality; and so has restored my sense of unity with nature.

From boyhood, I was deeply impressed by Wordsworth's lines in *Tintern Abbey:*

And I have felt
A presence that disturbs me with the joy
Of elevated thoughts; a sense sublime
Of something far more deeply interfused,

Whose dwelling is the light of setting suns,
And the round ocean, and the living air,
And the blue sky, and in the mind of man.

Yet I was unable to see how experiences of this kind, though I could personally testify to their value, could be linked up with the framework of ideas that I was attempting to build up on the basis of my scientific education. In the light of evolutionary humanism, however, the connection became clear, though the intellectual formulation given to it by Wordsworth was inadequate. The reality behind his thought is that man's mind is a partner with nature: it participates with the external world in the process of generating awareness and creating values.

The importance of this idea of participating, of co-operative partnership in a joint enterprise, had been brought home to me in various separate contexts. I had met with it as a keystone of our colonial policy in Africa; as a necessary basis for the work of UNESCO; as the concept inspiring the Colombo Plan and the United Nations' programme of technical assistance; as the basis for Bertrand de Jouvenel's illuminating definition of politics as action directed towards inducing men to co-operate in a common enterprise; indeed, evolutionary biology showed me the destiny of man on earth as a partnership between man and nature, with man in the leading position—a joint enterprise involving the participation of the entire human species for its most fruitful execution.

Finally, the concept of evolutionary humanism has helped me to see how, in principle at least, science and religion can be reconciled. It has shown me outlets for ideas and sentiments which can legitimately be called religious, but which otherwise would have remained frustrated or untapped. And it has indicated how vital a contribution science can make to religious progress.

My grandfather, in the same famous essay in which he defined agnosticism, stated as self-evident that "every man should be able to give a reason for the faith that is in him". My faith is in human possibilities: I hope that I have here succeeded in making clear some of my reasons for that faith.

———

This day before dawn I ascended a hill, and look'd at the
    crowded heaven,
And I said to my Spirit, *When we become the enfolders of
    those orbs, and the pleasure and knowledge of everything
    in them, shall we be fill'd and satisfied then?*
And my Spirit said, No, *we but level that lift, to pass and
    continue beyond.*

—*Walt Whitman*

Not all biologists responded enthusiastically as did Julian Huxley to Teilhard's blend of science and spirituality. George Gaylord Simpson (who, earlier in this book, sparred with Huxley on the question of progress in evolution) criticized The Phenomenon of Man *in a review published in* Scientific American *in 1960. Simpson judged that* The Phenomenon of Man *"provides a fascinating glimpse into the mind of a great soul, a kindly man and a subtle mystic. It may prove to be psychologically and historically important if, as is quite possible, it eventuates in a new religious cult of mystical evolutionism. It may do good (but could also conceivably do harm) in forcing theologians to face the fact of evolution more squarely. Despite its own claims and those of some of its sponsors and reviewers, it should not be taken either as a scientific treatise on evolution or as a derivation of religious conclusions from scientific premises."*

Teilhard's Toughest Critic

British biologist Peter Medawar (1915–1987) lodged perhaps the harshest criticism of Teilhard's worldview.

Peter Medawar

[*The Phenomenon of Man*] is a book widely held to be of the utmost profundity and significance; it created something like a sensation upon its publication a few years ago in France, and some reviewers hereabouts have called it the Book of the Year—one, the Book of the Century. Yet the greater part of it is nonsense, tricked out by a variety of tedious metaphysical conceits, and its author can be excused of dishonesty only on the grounds that before deceiving others he has taken great pains to deceive himself.

Medawar, who shared a Nobel Prize with Macfarlane Burnet for fundamental work on tissue grafting and "acquired immunological tolerance," published his critique in a 1961 issue of the journal Mind *(70:99–106). Excerpts here (copyright 1961 by Oxford University Press) are reprinted with permission of Mrs. Peter (Jean) Medawar and courtesy of Oxford University Press.*

   Medawar begins his review with a point-by-point refutation of some of Teilhard's main ideas. One he judged to be so fatuous that merely "to expound is to expose." He continues:

What Teilhard seems to be trying to say is that evolution is often (he says always) accompanied by an increase of orderliness or internal coherence or degree of integration. In what sense is the fertilized egg that develops into an adult human being 'higher' than, say, a bacterial cell? In the sense that it contains richer and more complicated genetical instructions for the execution of those processes that together constitute development. Thus Teilhard's radial, spiritual or psychic energy may be equated to 'information' or 'information content' in the sense that has been made reasonably precise by modern communications engineers. To equate it to consciousness, or to regard degree of consciousness as a measure of information content, is one of the silly little metaphysical conceits I mentioned in an earlier paragraph. Teilhard's belief, enthusiastically shared by Sir Julian Huxley, that evolution flouts or foils the second law of thermodynamics is based on a confusion of thought; and the idea that evolution has a main track or privileged axis is unsupported by scientific evidence.

Medawar concludes his critique with speculation on why, given its scientific flaws, The Phenomenon of Man *captured the attention of the intellectual world.*

How have people come to be taken in by *The Phenomenon of Man?* We must not underestimate the size of the market for works of this kind, for philosophy-fiction. Just as compulsory primary education created a market catered for by cheap dailies and weeklies, so the spread of secondary and latterly of tertiary education has created a large population of people, often with well developed literary and scholarly tastes, who have been educated far beyond their capacity to undertake analytical thought. It is through their eyes that we must attempt to see the attractions of Teilhard, which I shall jot down in the order in which they come to mind.

1. *The Phenomenon of Man* is anti-scientific in temper (scientists are shown up as shallow folk skating about on the surface of things), and, as if that were not recommendation enough, it was written by a scientist, a fact which seems to give it particular authority and weight. Laymen firmly believe that scientists are one species of person. They are not to know that the different branches of science require very different aptitudes and degrees of skill for their prosecution. Teilhard practised an intellectually unexacting kind of science in which he achieved a moderate proficiency. He has no grasp of what makes a logical argument or of what makes for proof. He does not even preserve the common decencies of scientific writing, though his book is professedly a scientific treatise.

2. It is written in an all but totally unintelligible style, and this is construed as prima facie evidence of profundity. It is because Teilhard has such wonderful deep thoughts that he's so difficult to follow—really, it's beyond my poor brain but doesn't that just show how profound and important it must be?

3. It declares that Man is in a sorry state, the victim of a "fundamental anguish of being", a "malady of space-time", a sickness of "cosmic gravity". The Predicament of Man is all the rage now that people have sufficient leisure and are sufficiently well fed to contemplate it, and many a tidy little literary reputation has been built upon exploiting it; anybody nowadays who dared to suggest that the plight of man might not be wholly desperate would get a sharp rap over the knuckles in any literary weekly. Teilhard not only diagnoses in everyone the fashionable disease but propounds a remedy for it—yet a remedy so obscure and so remote from the possibility of application that it is not likely to deprive any practitioner of a living.

4. *The Phenomenon of Man* was introduced to the English-speaking world by Sir Julian Huxley, which seemed to give it scientific benediction. Unlike myself, Sir Julian finds Teilhard in possession of a "rigorous sense of values", one who "always endeavoured to think concretely". He was speculative, to be sure, but his speculation was "always disciplined by logic". The only common ground between us is that Huxley, too, finds Teilhard somewhat difficult to follow. But then it does not seem to me that Huxley expounds Teilhard's argument; his Introduction does little more than to call attention to parallels between Teilhard's thinking and his own. Chief among these is the cosmic significance attached to a suitably generalized conception of evolution—a conception so diluted or attenuated in the course of being generalized as to cover all events or phenomena that are not immobile in time. In particular, Huxley applauds the, in my opinion, superficial and ill thought out view that the so-called 'psycho-social evolution' of mankind and the genetical evolution of living organisms generally are two episodes of a continuous integral process (though separated by a "critical point", whatever that may mean). Yet for all this Huxley finds it impossible to follow Teilhard "all the way in his gallant attempt to reconcile the supernatural elements of Christianity with the facts and implications of evolution". But, bless my soul, this reconciliation is just what Teilhard's book is about! And so, it seems to me, Huxley contrives to enrage all parties—those who have some concern for rigorous analytical thought, and those who see in Teilhard's work the elements of a profound spiritual revelation.

I have read and studied *The Phenomenon of Man* with real distress, even with despair. Instead of wringing our hands over the Human Predicament, we should attend to those parts of it which are wholly remediable, above all to the gullibility which makes it possible for people to be taken in by such a bag of tricks as this. If it were an innocent, passive gullibility it would be excusable; but all too clearly, alas, it is an active willingness to be deceived.

———

Whole droves of minds are, by the driving god,
Compelled to drink the deep Lethaean flood:

In large forgetful draughts to steep the cares
Of their past labors, and their irksome years.

—*Voltaire*

Chance and Necessity

A decade after The Phenomenon of Man *appeared in English, Jacques Monod produced its antipode.* Chance and Necessity: An Essay on the Natural Philosophy of Modern Biology *has become a scientific classic. Monod proves that a stark and (to some) unsavory view of the cosmos does not demand a stark and unsavory presentation.* Chance and Necessity *combines literary elegance with the authority of a Nobel laureate.*

Jacques Monod (1910–1976) shared a 1965 Nobel Prize with André Lwoff and François Jacob for discoveries in molecular biology that today form the basis of our understanding of how genetic information is actively expressed by way of protein synthesis. Like Teilhard, Monod was a citizen of France, but there the similarities end. The worldviews of Teilhard and Monod are so discordant that a reader's enchantment with one is almost surely to be matched by disdain for the other. Monod summarizes his message thus:

Jacques Monod

The thesis I shall present in this book is that the biosphere does not contain a predictable class of objects or of events but constitutes a particular occurrence, compatible indeed with first principles, but not deducible from those principles and therefore essentially unpredictable. Let there be no misunderstanding here. In saying that as a class living beings are not predictable upon the basis of first principles, I by no means intend to suggest that they are not explicable through these principles—that they transcend them in some way, and that other principles, applicable to living systems alone, must be invoked. In my view the biosphere is unpredictable for the very same reason—neither more nor less—that the particular configuration of atoms constituting this pebble I have in my hand is unpredictable. No one will find fault with a universal theory for not affirming and foreseeing the existence of this particular configuration of atoms; it is enough for us that this actual object, unique and real, be compatible with the theory. This object, according to the theory, is under no obligation to exist; but it has the right to.

That is enough for us as concerns the pebble, but not as concerns ourselves. We would like to think ourselves necessary, inevitable, ordained from all eternity. All religions, nearly all philosophies, and even a part of science testify to the unwearying, heroic effort of mankind desperately denying its own contingency.

Why did Monod write Chance and Necessity?

Nowadays a man of science is not well advised to use the word 'philosophy', albeit qualified as 'natural', in the title (or even the subtitle) of a book: nothing more is needed to earn it a distrustful reception from other scientists, and from philosophers a condescending one at best. I have only one excuse, but I believe it is sound: the duty which more forcibly than ever thrusts itself upon scientists to apprehend their discipline within the larger framework of modern culture, with a view to enriching the latter not only with technically important findings, but also with what they may feel to be humanly significant ideas arising from their area of special concern. The very ingenuousness of a fresh look at things (and science possesses an ever-youthful eye) may sometimes shed a new light upon old problems. Meanwhile, to be sure, any confusion between the ideas suggested by science and science itself must be carefully avoided; but it is just as necessary that scientifically warranted conclusions be resolutely pursued to the point where their full meaning becomes clear.

Monod wrote this book, in part, to contest Teilhard de Chardin's optimistic and purposeful view of evolution, which had won a considerable following in the public at large. Six years later, upon Monod's death, Francis Crick assessed the impact of Chance and Necessity.

Francis Crick

It aroused the almost united opposition of the French intellectual establishment, which has always preferred Marx, Freud and Teilhard de Chardin to Darwin and Mendel. Written with force and clarity, in an unmistakable personal style, it presented a view of the universe that to many lay readers appeared strange, sombre, arid and austere. This is all the more surprising since the central vision of life that it projected is shared by the great majority of working scientists of any distinction. It would be difficult to find a better example to display the deep rift between science and the rest of our culture.

Selections from Chance and Necessity *by Jacques Monod, translated by Austryn Wainhouse (copyright 1971 by Alfred A. Knopf, Inc.), are reprinted by permission of Alfred A. Knopf, Inc. and by Collins, and imprint of HarperCollins Publishers, Ltd.*

The Animist Covenant

Jacques Monod

Our ancestors, we must presume, perceived the strangeness of their condition only dimly. They did not have the reasons we have today for feeling themselves strangers in the universe upon which they opened their eyes. What did they see first? Animals, plants; beings whose nature they could at once divine as similar to their own. Plants grow, seek sunlight, die; animals stalk their prey, attack their enemies, feed and protect their young; males fight for the possession of a female. About plants and ani-

Native American rock art. *Upper left:* petroglyph, Wyoming. *Lower left:* red paint, Texas. *Upper right:* red paint, Ontario. *Lower right:* petroglyph, California.

mals as about man himself there was nothing hard to explain. These beings all have an aim, a purpose: to live and to go on living in their progeny, even at the price of death. Its purpose explains the being, and the being makes sense only through the purpose animating it.

But around them our ancestors also saw other objects, far more mysterious: rocks, rivers, mountains, the thunderstorm, the rain, the stars in the sky. If these objects exist it must also be for a purpose; to nourish it they had also to have a spirit or soul. Thus was the world's strangeness resolved for those early human beings: in reality there exist no inanimate objects. For such a thing would be incomprehensible. In the river's depths, on the mountaintop, more subtle spirits pursue vaster and more impenetrable designs than the transparent ones animating men and beasts. Thus were our forebears wont to see in nature's forms and events the action of forces either benign or hostile, but never indifferent—never totally alien.

Animist belief, as I am visualizing it here, consists essentially in a projection into inanimate nature of man's awareness of the intensely teleonomic *(endowed with a purpose or project)* functioning of his own central nervous system. It is, in other words, the hypothesis that natural phenomena can and must be explained in the same manner, by the same "laws," as subjective human activity, conscious and purposive. Primitive animism formulated this hypothesis with complete candor, frankness, and precision, populating nature with gracious or awesome myths and myth-figures which have for centuries nourished art and poetry.

One would be wrong to smile, even out of the fondness and deference the childlike inspire. Do we imagine that modern culture has really forsaken the subjective interpretation of nature? Animism established a covenant between nature and man, a profound alliance outside of which seems to stretch only terrifying solitude. Must we break this tie because the postulate of objectivity requires it?

Ever since the seventeenth century the history of ideas attests to the profuse efforts put forth by the greatest minds to avert that break, to forge the old bond anew. Think of such mighty efforts as those of Leibnitz, or of the colossal and ponderous monument Hegel raised. But idealism has not by any means been the only refuge for a cosmic animism. At the very core of certain ideologies said and claiming to be founded upon science, the animist projection, in a more or less disguised form, turns up again.

The biological philosophy of Teilhard de Chardin would not merit attention but for the startling success it has encountered even in scientific circles. A success which tells of the eagerness, of the need to revive the covenant. Teilhard revives it, and does so nakedly.

Although Teilhard's logic is hazy and his style laborious, some of those who do not entirely accept his ideology yet allow it a certain poetic grandeur. For my part I am most of all struck by the intellectual spinelessness

of the philosophy. In it I see more than anything else a systematic truckling, a willingness to conciliate at any price, to come to any compromise. Perhaps, after all, Teilhard was not for nothing a member of that order which, three centuries earlier, Pascal assailed for its theological laxness.

The idea of re-establishing the old animist covenant with nature, or of founding a new one through a universal theory according to which the evolution of the biosphere culminating in man would be part of the smooth onward flow of cosmic evolution itself—this idea did not of course originate with Teilhard. It is in fact the central theme of nineteenth-century scientistic progressism. One finds it at the very heart of Spencer's positivism and of the dialectical materialism of Marx and Engels as well. The unknown and unknowable force which, according to [Herbert] Spencer, operates throughout the universe creating variety, coherence, specialization, and order, plays what amounts to exactly the same role in Teilhard's "ascending" energy: human history is the extension of biological evolution, itself a component part of cosmic evolution. Thanks to this single principle, man at last finds his eminent and necessary place in the universe, along with certainty of the progress which is forever pledged to him.

Spencer's differentiating force, like Teilhard's ascending energy, is a plain instance of animist projection. In order to give meaning to nature, so that man need not be separated from it by a fathomless gulf, and for it again to become decipherable and intelligible, a purpose had to be restored to it. Should no spirit be available to harbor this purpose, then one inserts into nature an evolutive, an ascending "force," which in effect amounts to abandoning the postulate of objectivity.

Platonic and Heraclitean Roots of Modern Biology

Ever since its birth in the Ionian Islands almost three thousand years ago, Western philosophy has been divided between two seemingly opposed attitudes. According to one of them the authentic and ultimate truth of the world can reside only in perfectly immutable forms, by essence unvarying. According to the other, the only real truth resides in flux and evolution. From Plato to Whitehead and from Heraclitus to Hegel and Marx, it is clear that these metaphysical epistemologies were always closely bound up with their authors' ethical and political biases. These ideological edifices, represented as self-evident to reason, were actually *a posteriori* constructions designed to justify preconceived ethico-political theories.

For science the only *a priori* is the postulate of objectivity, which spares—or rather forbids—it from taking part in the debate. Science studies evolution, whether of the universe or of the systems it contains, such as the biosphere, including man. We are aware that any phenomenon, any event, any cognition implies interactions which by themselves generate modifications in the elements of the system. From this it does not follow that the existence of immutable entities within the structure of the universe must be denied. Quite the contrary: for the basic strategy of science in the analysis of phenomena is the ferreting out of invariants. In science there is and will remain a Platonic element which could not be taken away without ruining it. Amidst the infinite diversity of singular phenomena, science can only look for invariants.

A blue alga, an infusorian, an octopus, and a human being—what have they in common? With the discovery of the cell and the advent of cellular theory a new unity could be seen under this diversity. But it was some time before advances in biochemistry, mainly during the second quarter of this century, revealed the profound and strict oneness, on the microscopic level, of the whole of the living world. Today we know that from the bacterium to man the chemical machinery is essentially the same, in both its structure and its functioning.

In its structure: all living beings, without exception, are made up of the same two principal classes of macromolecular components: proteins and nucleic acids. What is more, these macromolecules are in all living beings constituted by the assembling of the same residues, finite in number: twenty amino acids for the proteins and four kinds of nucleotides for the nucleic acids. In its functioning: the same reactions, or rather sequences of reactions, are used in all organisms for the essential chemical operations: the mobilization and storing of chemical potential, the biosynthesis of cellular components.

True, upon this central theme of metabolism many variations are to be met with, each corresponding to a particular functional adaptation. However, they almost always consist in new utilizations of universal metabolic sequences, hitherto employed for other functions. For instance, the excretion of nitrogen occurs in different forms in birds and mammals: the former excrete uric acid, the latter urea. Now the pathway for the synthesis of uric acid in birds is only a modification, a minor one moreover, of the sequence of reactions which in all organisms synthesizes the so-called purine nucleotides, universal components of nucleic acids. In mammals the synthesis of urea is obtained thanks to a modification of another universal metabolic pathway: the one which concludes with the synthesis of arginine, an amino acid present in all proteins. Any number of examples could be given.

To biologists of my generation fell the discovery of the virtual identity of cellular chemistry throughout the entire biosphere. By 1950 research pointed to it as a certainty, and each new publication added further confirmation of it. The hopes of the most convinced "Platonists" were being more than gratified. But this gradual disclosure of the universal "form" of cellular chemistry seemed, in the meantime, to render the problem of reproductive invariance still more acute and more paradoxical. If, chemically, the components are the same and are synthesized by the same processes in all living beings, what is the source of their prodigious morphological and physiological diversity? And, yet more puzzling, how does each species, using the same materials and the same chemical transformations as all the others, maintain, unchanged from generation to generation, the structural standard that characterizes it and differentiates it from every other?

We now have the solution to this problem. The universal components—the nucleotides on the one side, the amino acids on the other—are the logical equivalents of an alphabet in which the structure and consequently the specific associative functions of proteins are spelled out. In this alphabet can therefore be written all the diversity of structures and performances the biosphere contains. More, with each succeeding cellular generation it is the reproduction of the text, written in the form of DNA nucleotide sequences, that guarantees the invariance of the species.

The fundamental biological invariant is DNA. That is why Mendel's defining of the gene as the unvarying bearer of hereditary traits, its chemical identification by Avery (confirmed by Hershey), and the elucidation by Watson and Crick of the structural basis of its replicative invariance, without any doubt constitute the most important discoveries ever made in biology. To which of course must be added the theory of natural selection, whose certainty and full significance were established only by those later discoveries.

Physics tells us however that—save at absolute zero, an inaccessible limit—no microscopic entity can fail to undergo quantum perturbations, whose accumulation within a macroscopic system will slowly but surely alter its structure. Living beings, despite the perfection of the machinery that guarantees the faithfulness of the translation *(of the DNA genetic code into proteins for building and repairing the body)*, are not exempt from this law. Aging and death in pluricellular organisms is accounted for, at least in part, by the piling up of accidental errors of translation. These, in particular affecting certain components responsible for the accuracy of translation, tend to precipitate further errors which, ever more frequent, gradually and inexorably undermine the structure of those organisms. Nor, without violating the laws of physics, could the mechanism of replication be completely immune to disturbances, accident-proof. At least some of these disturbances create more or less discrete modifications in certain elements of the DNA sequence.

Chance as the Source of Innovation

Various mutations have been identified as due to: (1) the substitution of a single pair of nucleotides for another pair; (2) the deletion or addition of one or several pairs of nucleotides; and (3) various kinds of "scrambling"

Elément méchanique I
(Fernand Leger)

of the genetic text by inversion, duplication, displacement, or fusion of more or less extended segments. We call these events accidental; we say that they are random occurrences. And since they constitute the only possible source of modifications in the genetic text, itself the sole repository of the organism's hereditary structures, it necessarily follows that chance alone is at the source of every innovation, of all creation in the biosphere.

Pure chance, absolutely free but blind, at the very root of the stupendous edifice of evolution: this central concept of modern biology is no longer one among other possible or even conceivable hypotheses. It is today the sole conceivable hypothesis, the only one that squares with observed and tested fact. And nothing warrants the supposition—or the hope—that on this score our position is likely ever to be revised.

There is no scientific concept, in any of the sciences, more destructive of anthropocentrism than this one, and no other so rouses an instinctive protest from the intensely teleonomic creatures that we are. For every vitalist of animist ideology it is therefore the concept or rather the specter to be exorcized at all costs. And so it is most important to say something about the words 'chance' and 'randomness', and to specify in just what sense they may and must be used with regard to mutations as the source of evolution.

The idea of chance is not a simple one, and the word itself is employed in a wide variety of contexts. The best thing is to take a few examples. Dice or roulette are termed games of chance, and the theory of probability is used to forecast their outcome. But chance enters into these purely mechanical and macroscopic games only because of the practical impossibility of governing the throw of the dice or the spinning of the little ball with sufficient precision. An exceedingly precise mechanical thrower could conceivably be invented which would go far to reduce the uncertainty of the outcome. Let us say that in roulette the uncertainty is purely operational and not essential. The same holds, one will quickly see, for the theory of numerous phenomena where the concept of chance and the theory of probability are used for purely methodological reasons.

But in other situations the idea of chance takes on an essential and no longer merely operational meaning. Such is the case, for instance, in what may be called "absolute coincidences," those, that is to say, which result from the intersection of two totally independent chains of events. Suppose that Dr. Brown sets out on an emergency call to a new patient. In the meantime Jones the contractor's man has started making emergency repairs on the roof of a nearby building. As Dr. Brown walks past the building, Jones inadvertently lets go of his hammer, whose (deterministic) trajectory happens to intercept that of the physician, who dies of a fractured skull. We say he was a victim of chance. What other term fits such an event, by its very nature unforeseeable? Chance is obviously the essential thing here, inherent in the complete independence of two causal chains of events whose convergence produces the accident.

Now, between the occurrences that can provoke or permit an error in the replication of the genetic message and its functional consequences there is also a complete independence. The functional effect depends upon the structure, upon the actual role of the modified protein, upon the interactions it ensures, upon the reactions it catalyzes—all things which have nothing to do with the mutational event itself nor with its immediate or remote causes, regardless of the nature, whether deterministic or not, of those causes.

Finally, on the microscopic level there exists a further source of still more radical uncertainty, embedded in the quantum structure of matter. A mutation is in itself a microscopic event, a quantum event, to which the principle of uncertainty consequently applies. An event which is hence and by its very nature essentially unpredictable.

The principle of uncertainty was never entirely accepted by some of the greatest modern physicists, Einstein foremost among them, who was unwilling to admit that "God plays at dice." Certain schools have retained it for its operational usefulness but denied it the standing of an essential concept. However, all the efforts made to replace quantum theory by a "finer" structure from which uncertainty has vanished have ended in failure, and at the present time *(still true twenty years after Monod wrote this)* very few physicists seem disposed to believe that this principle will ever disappear from their discipline.

At any rate it must be stressed that, even were the principle of uncertainty someday abandoned, it would remain nonetheless true that between the determination, however complete, of a mutation of DNA and the determination of its functional effects on the plane of protein interaction one could still see nothing but an "absolute coincidence" like that defined above by the parable of the workman and the physician. The occurrence would still belong to the realm of "essential" chance.

Necessity and 'Choice' in Natural Selection

The initial elementary events which open the way to evolution in the intensely conservative systems called living beings are microscopic, fortuitous, and utterly without relation to whatever may be their effects upon teleonomic functioning. But once incorporated in the DNA structure, the accident—essentially unpredictable because always singular—will be mechanically and faithfully replicated and translated: that is to say, both multiplied and transposed into millions or billions of copies. Drawn out of the realm of pure chance, the accident enters into that of necessity, of the most implacable certainties. For natural selection operates at the macroscopic level, the level of organisms.

Even today a good many distinguished minds seem unable to accept or even to understand that from a source of noise natural selection alone and unaided could have drawn all the music of the biosphere. In effect natural selection operates upon the products of chance and can feed nowhere else; but it operates in a domain of very demanding conditions, and from this domain chance is barred. It is not to chance but to these conditions that evolution owes its generally progressive course, its successive conquests, and the impression it gives of a smooth and steady unfolding.

The extraordinary stability of certain species, the billions of years spanned by evolution, the invariance of the cell's basic chemical scheme—these obviously can be explained only by the extreme coherence of the teleonomic system which in evolution has acted as both guide and brake, and has retained, amplified, and integrated only a tiny fraction of the myriad opportunities offered it by nature's roulette. For its part the replicative system, far from being able to eliminate the microscopic perturbations by which it is inevitably beset, knows only how to register and offer them—almost always in vain—to the teleonomic filter by which their performance is finally judged, through natural selection.

The selective pressures exerted by outside conditions upon organisms are in no case unconnected with the teleonomic performances characteristic of the species. Different organisms inhabiting the same ecological niche interact in very different and specific ways with outside conditions (among which one must include other organisms). These specific interactions, which the organism itself "elects," at least in part, determine the nature and orientation of the selective pressure the organism sustains. Let us say that the "initial conditions" of selection encountered by a new mutation simultaneously and inseparably include both the environment surrounding the organism and the total structures and performances of the teleonomic apparatus belonging to it.

Obviously, the part played by teleonomic performances in the orientation of selection becomes greater and greater, the higher the level of organization and hence autonomy of the organism with respect to its environment—to the point where teleonomic performance may indeed be considered decisive in the higher organisms, whose survival and reproduction depend above all upon their behavior. It is evident as well that the initial choice of this or that kind of behavior can often have very long-range consequences, affecting not only the species in which it first crops up in rudimentary form, but all its descendants, even if these should constitute an entire evolutionary subgroup.

As we all know, the great turning points in evolution have coincided with the invasion of new ecological spaces. If terrestrial vertebrates appeared and were able to initiate that wonderful line from which amphibians, reptiles, birds, and mammals later developed, it was originally because a primitive fish "chose" to do some exploring on land, where it was however ill-provided with means for getting about. The same fish thereby created, as a consequence of a shift in behavior, the selective pressure which was to engender the powerful limbs of the quadrupeds. Among the descendants of this daring explorer, this Magellan of evolution, are some that can run at speeds of fifty miles an hour; others climb trees with astonishing agility, while yet others have conquered the air, in a fantastic manner fulfilling, extending, and amplifying the ancestral fish's hankering, its "dream."

The fact that in the evolution of certain groups one observes a general tendency, maintained over millions of years, toward the apparently oriented development of certain organs shows how the initial choice of a

Mudskippers (*Periophthalmus* or *Periophthalmodon*) on mangrove roots.

certain kind of behavior (for example, in the face of attack from a predator) commits the species irrevocably in the direction of a continuous perfecting of the structures and performances this behavior needs for its support. It is because the ancestors of the horse at an early point chose to live upon open plains and to flee at the approach of an enemy (rather than try to put up a fight or hide) that the modern species, following a long evolution made up of many stages of reduction, today walks on the tip of a single toe.

Some post-Darwinian evolutionists have tended, when they discussed natural selection, to propagate a stark, naively ferocious idea of it: that of the no-holds-barred "struggle for life"—an expression which comes not from Darwin but from Herbert Spencer. The neo-Darwinians of the beginning of this century for their part proposed a much richer concept and showed, on the basis of quantitative theories, that the decisive factor in natural selection is not the struggle for life, but —within a species—the differential rate of reproduction.

It is known that certain very precise and complex kinds of behavior, such as the prenuptial ceremonies of birds, are narrowly linked to certain especially conspicuous morphological features. There can be no doubt that this behavior and the anatomical particularities that go with it evolved *pari passu,* each encouraging and reinforcing the other under the pressure of sexual selection. Once it starts to develop in a species, any decorative finery associated with successful mating only adds to, in short confirms, the initial pressure of selection and consequently favors any improvement in the finery itself. One is therefore quite right in saying that the sexual drive—or better still, desire—created the conditions under which some magnificent plumages were selected.

Lamarck was of the belief that the very strain entailed in an animal's efforts to "succeed" in life somehow affected its hereditary legacy, entering into it and having a direct modeling influence upon its descendants. The giraffe's immensely long neck would thus express its forebears' unabating wish to reach the topmost branches of trees. This is of course today an unacceptable hypothesis; yet one sees that pure selection, operating upon elements of behavior, leads to the result Lamarck sought to explain: the close interconnection of anatomical adaptations and specific performances.

The Origin of Life and of Consciousness

When one ponders on the tremendous journey of evolution over the past three billion years or so, the prodigious wealth of structures it has engendered, and the extraordinarily effective teleonomic performances of living beings, from bacteria to man, one may well find oneself beginning to doubt again whether all this could conceivably be the product of an enormous lottery presided over by natural selection, blindly picking the rare winners from among numbers drawn at utter random. While one's conviction may be restored by a detailed review of the accumulated modern evidence that this conception alone is compatible with the facts (notably with the molecular mechanisms of replication, mutation, and translation), it affords no synthetic, intuitive, and immediate grasp of the vast sweep of evolution. The miracle stands "explained"; it does not strike us as any less miraculous. As François Mauriac wrote, "What this professor says is far more incredible than what we poor Christians believe."

This is true, just as it is true that there is no achieving a satisfactory mental image of certain abstractions in modern physics. But we also know that such difficulties can not be taken as arguments against a theory which is vouched for by experiment and logic. In the case of physics, microscopic or cosmological, we see right away what the trouble is: the scale of the envisaged phenomena transcends the categories of our immediate experience. Only abstraction can supply this deficiency, yet without curing it. Where biology is concerned the difficulty is of another order. The elementary interactions upon which everything hinges, these, thanks to their "mechanical" character, are relatively easy to grasp. Much less readily come by is an intuitive global picture of living systems whose phenomenal complexity defies assimilation. But in biology as in physics these psychological difficulties, once again, do not constitute an argument against theory and observation.

Central to biology, the evolutionary concept is bound to undergo considerable elaboration in the years ahead. Much remains to be learned. Essentially, however, the problem has been resolved and evolution now lies well to this side of the frontier of knowledge. The present challenge, as I see it, is in the areas at the two extremes of evolution: the origin of the first living systems on the one hand; on the other, the inner workings of the most intensely teleonomic system ever to have emerged, to wit, the central nervous system of Man.

Life appeared on earth: what, before the event, were the chances that this would occur? The present structure of the biosphere far from excludes the possibility that the decisive event occurred only once. Which would mean that its a priori probability was virtually zero. This idea is distasteful to most scientists. Science can neither say nor do anything about a unique occurrence. It can only consider occurrences that form a class, whose a priori probability, however faint, is yet definite. Now through the very universality of its structures, starting with the [genetic] code, the biosphere looks like the product of a unique event. It is possible of course that its uniform character was arrived at by elimination through selection of many other attempts or variants. But nothing compels this interpretation. At the present time we have no legitimate grounds for either asserting or denying that life got off to but a single start on earth, and that, as a consequence, before it appeared its chances of occurring were next to nil.

Not only for scientific reasons do biologists recoil at this idea. It runs counter to our very human tendency to believe that behind everything real in the world stands a necessity rooted in the very beginning of things. Against this notion, this powerful feeling of destiny, we must be constantly on guard.

Immanence is alien to modern science. Destiny is written concurrently with the event, not prior to it. Our own was not written before the emergence of the human species, alone in all the biosphere to utilize a logical system of symbolic communication. Another unique event, which by itself should predispose us against any anthropocentrism. If it was unique, as may perhaps have been the appearance of life itself, then before it did appear its chances of doing so were infinitely slender. The universe was not pregnant with life nor the biosphere with man. Our number came up in the Monte Carlo game. Is it any wonder if, like the person who has just made a million at the casino, we feel strange and a little unreal?

Evolutionary Roots of Myths and Religions

During entire aeons a man's lot was identical with that of the group, of the tribe he belonged to and outside of which he could not survive. The tribe, for its part, was able to survive and defend itself only through its cohesion. Whence the extreme force of inward coercion exerted by the laws that organized and guaranteed this cohesion. A man might perhaps infringe them; it is not likely that any man ever dreamed of denying them. Given the immense selective importance such social structures perforce assumed over such vast stretches of time, it is difficult not to believe that they must have made themselves felt upon the genetic evolution of the innate categories of the human brain. This evolution must not only have facilitated acceptance of the tribal law, but created the need for the mythical explanation which gave it foundation and sovereignty.

We are the descendants of such men. From them we have probably inherited our need for an explanation, the profound disquiet which goads us to search out the meaning of existence. That same disquiet has created all the myths, all the religions, all the philosophies, and science itself.

That this imperious need develops spontaneously, that it is inborn, inscribed somewhere in the genetic code, strikes me as beyond doubt. Outside the human species, nowhere in the animal kingdom does one find such highly differentiated social organizations save it be among certain insects: ants, termites, bees. But the stability of the social insects' institutions owes next to nothing to cultural heritage, virtually everything to genetic transmission. Social behavior, with them, is entirely innate, automatic.

Man's social institutions, purely cultural, cannot ever attain a like stability; besides, who would wish for such a thing? The invention of myths and religions, the construction of vast philosophical systems—they are the price this social animal has had to pay in order to survive without having to yield to pure automatism. But to anchor the social structure, the cultural tradition, all by itself, would not have been reliable enough, strong enough. That heritage needed a genetic support to make it into something the mind could not do without. How else account for the fact that in our species the religious phenomenon is invariably at the base of social structure? How else explain that, throughout the immense variety of our myths, our religions, and philosophical ideologies, the same essential "form" always recurs?

It is readily seen that the "explanations" meant to give foundation to the law while assuaging man's anxiety are all narrations of past events, "stories"—or "histories"—that are (in the philosophical sense) really "ontogenies." Primitive myths almost all tell of more or less divine heroes whose deeds explain the origins of the group and base its social structure upon immutable traditions; one does not remake history. The great religions are of similar form, resting on the story of the life of an inspired prophet who, if not himself the founder of all things, represents that founder, speaks for him, and recounts the history of mankind as well as its destiny. Of all the great religions Judeo-Christianity is probably the most "primitive," since its strictly historicist structure is directly plotted upon the saga of a Bedouin tribe before being enriched by a divine prophet. Buddhism, on the contrary, more highly differentiated, has recourse in its original form to Karma alone, the transcending law governing individual destiny. Buddhism is more a story of souls than of men.

If it is true that the need for a complete explanation is innate, that its absence begets a profound ache within; if the only form of explanation capable of putting the soul at ease is that of a total history which discloses the meaning of man by assigning him a necessary place in nature's scheme; if, to appear genuine, meaningful, soothing, the "explanation" must blend into the long animist tradition, then we understand why it took so many thousands of years for the kingdom of ideas to be invaded by the one according to which objective knowledge is the only authentic source of truth. Cold and austere, proposing no explanation but imposing an ascetic renunciation of all other spiritual fare, this idea was not of a kind to allay anxiety, but aggravated it instead. By a single stroke it claimed to sweep away the tradition of a hundred thousand years, which had become one with human nature itself. It wrote an end to the ancient animist covenant between man and nature, leaving nothing in place of that precious bond but an anxious quest in a frozen universe of solitude. With nothing to recommend it but a certain puritan arrogance, how could such an idea win acceptance? It did not; it still has not. It has however commanded recognition; but that is because, solely because, of its prodigious power of performance.

The Estrangement of Science

In the course of three centuries science, founded upon the postulate of objectivity, has conquered its place in society—in men's practice, but not in their hearts. Modern societies are built upon science. They owe it their wealth, their power, and the certitude that tomorrow far greater wealth and power still will be ours if we so wish. But there is this too: just as an initial "choice" in the biological evolution of a species can be binding upon its entire future, so the choice of a scientific practice, an unconscious choice in the beginning, has launched the evolution of culture on a one-way path; onto a track which nineteenth-century scientism saw lead-

judge of them and must ignore them; but it subverts every one of the mythical or philosophical ontogenies upon which the animist tradition, from the Australian aborigines to the dialectical materialists, has made all ethics rest: values, duties, rights, prohibitions.

If he accepts this message—accepts all it contains—then man must at last wake out of his millenary dream; and in doing so, wake to his total solitude, his fundamental isolation. Now does he at last realize that, like a gypsy, he lives on the boundary of an alien world. A world that is deaf to his music, just as indifferent to his hopes as it is to his suffering or his crimes.

But henceforth who is to define crime? Who shall decide what is good and what is evil? All the traditional systems have placed ethics and values beyond man's reach. Values did not belong to him; he belonged to them. He now knows that they are his and his alone, and they no sooner come into his possession than lo! they seem to melt into the world's uncaring emptiness. It is then that modern man turns toward science, or rather against it, finally measuring its terrible capacity to destroy not only bodies but the soul itself.

Where is the remedy? Must one adopt the position once and for all that objective truth and the theory of values constitute eternally separate, mutually impenetrable domains? This is the attitude taken by a great number of modern thinkers, whether writers, or philosophers, or indeed scientists. For the vast majority of men, whose anxiety it can only perpetuate and worsen, this attitude I believe will not do. I also believe it is absolutely mistaken, and for two essential reasons. First, and obviously, because values and knowledge are always and necessarily associated in action just as in discourse. Second, and above all, because the very definition of "true" knowledge reposes in the final analysis upon an ethical postulate.

The Ethic of Knowledge

True knowledge is ignorant of values, but it cannot be grounded elsewhere than upon a value judgment, or rather upon an axiomatic value. It is obvious that the positing of the principle of objectivity as the condition of true knowledge constitutes an ethical choice and not a judgment arrived at from knowledge, since, according to the postulate's own terms, there cannot have been any "true" knowledge prior to this arbitrary choice. Thus, assenting to the principle of objectivity one announces one's adherence to the basic statement of an ethical system, one asserts the ethic of knowledge.

Hence it is from the ethical choice of a primary value that knowledge starts. The ethic of knowledge thereby differs radically from animist ethics, which all claim to be based upon the "knowledge" of immanent laws, religious or "natural," which are supposed to assert themselves over man. The ethic of knowledge does not obtrude itself upon man; on the contrary, it is he who prescribes it to himself, making of it the axiomatic condition of authenticity for all discourse and all action.

Authentic discourse in its turn lays the foundation of science, and returns to the hands of man the immense powers that enrich and imperil him today. Modern societies, woven together by science, living from its products, have become as dependent upon it as an addict on his drug. They owe their material wherewithal to this fundamental ethic upon which knowledge is based, and their moral weakness to those value-systems, devastated by knowledge itself, to which they still try to refer. The contradiction is deadly. It is what is digging the pit we see opening under our feet. The ethic of knowledge that created the modern world is the only ethic compatible with it, the only one capable, once understood and accepted, of guiding its evolution.

Understood and accepted—could it be? If it is true, as I believe, that the fear of solitude and the need for a complete and binding explanation are inborn—that this heritage from the very remote past is not only cultural but probably genetic too—can one imagine such an ethic as this, austere, abstract, proud, calming that fear, satisfying that need? I do not know. But it may not be altogether impossible. Perhaps, even more than an explanation which the ethic of knowledge cannot supply, it is to rise above himself that man craves. The abiding power of the great socialist dream, still present in men's hearts, would indeed seem to suggest it.

No system of values can be said to constitute a true ethic unless it proposes an ideal reaching beyond the individual and transcending the self to the point even of justifying self-sacrifice, if need be. By the very loftiness of its ambition the ethic of knowledge might perhaps satisfy this urge in man to project toward something higher. It sets forth a transcendent value, true knowledge, and invites him not to use it self-servingly but henceforth to enter into its service from deliberate and conscious choice. At the same time it is also a humanism, for in man it respects the creator and repository of that transcendence.

The ethic of knowledge is also in a sense "knowledge of ethics," a clear-sighted appreciation of the urges and passions, the requirements and limitations of the biological being. It is able to confront the animal in man, to view him not as absurd but strange, precious in his very strangeness: the creature who, belonging simultaneously to the animal kingdom and the kingdom of ideas, is simultaneously torn and enriched by this agonizing duality, alike expressed in art and poetry and in human love.

The ancient covenant is in pieces; man knows at last that he is alone in the universe's unfeeling immensity, out of which he emerged only by chance. His destiny is nowhere spelled out, nor is his duty. The kingdom above or the darkness below: it is for him to choose.

There was a time when meadow, grove, and stream,
The earth, and every common sight,
   To me did seem
   Apparelled in celestial light,
The glory and the freshness of a dream.
It is not now as it hath been of yore;—
   Turn wheresoe'er I may,
   By night or day,
The things which I have seen I now can see no more.

Whither is fled the visionary gleam?
Where is it now, the glory and the dream?

—*William Wordsworth*

As with Teilhard de Chardin, both criticism and praise of Jacques Monod's ideas remain spirited to this day. Of the scientists and philosophers of science who have found fault with Monod, comments of three will be highlighted here.

Finding Fault with Monod

Mary Midgley, a philosopher, derides the widespread appeal of Monod's "world-picture" in her 1985 book, Evolution as a Religion.

Mary Midgley

It is not surprising that Monod's story has had so much success, especially among scientists. In its lively, existentially coloured package, it offers a way of combining the general scepticism and acceptance of confusion about moral questions which is widely professed today with a firm, saving exception for confidence in the value of science. This fits the world-picture acquired by very many people in the course of a scientific education, an education which trains them in scientific thinking, and greatly exaggerates the precision possible to it, while doing very little to teach them the ways of thinking which they will need for other purposes—personal, political, psychological, historical, metaphysical and all the rest.

Since these purposes are central to life and call for a great deal of thought, especially in changing times, those whose intellect has been cramped by this kind of foot-binding process into a specialized use experience a very painful sense of confusion when other issues come before them. The discrepancy between their confident use of highly trained intelligence in their work and their helplessness on other issues threatens to tear them apart and attacks the roots of their self-respect. In this emergency, Monod appears with the balm of a metaphysical proof that their plight is inescapable. He declares that science is indeed the only field where thought is possible. Everything else must be left to choice: not reasonable choice, but choice in the Existentialist sense of a blind, inarticulate act of will.

Biologist Stanley Salthe, in a 1989 essay, judges Chance and Necessity *"masterful—and morally demonic." He goes on to say:*

Stanley Salthe

We are, as evolutionary biologists, indirectly working on nothing less than an important part of our culture's very own creation myth. Is the combination of the pointlessness of chance with the tyranny of necessity, competitive exclusion, expedience, and obedience to external forces what we really want to think of as the sources of our origins?

Theodosius Dobzhansky, who pioneered the use of fruitflies for understanding genetic mutation, pictures a world that is neither Monodian black nor Teilhardian white. The Dobzhansky sampler here is compiled from a variety of his works, listed in the bibiography.

Theodosius Dobzhansky

It is impossible to write books like Teilhard's or Monod's without going beyond the bounds of "objective knowledge", and revealing one's metaphysical, esthetic and emotional attitudes and sympathies. Teilhard's world view is frankly centred on his Christian faith. Monod's is just as clearly anti-Christian, and following the lead of French existentialists.

Evolutionary biology is pivotal in Teilhard's as well as in Monod's views. However, the two authors view the evolutionary process in quite different perspectives. The crucial issue turns out to be whether evolution is going in some direction, or if it is a senseless happenstance. Teilhard had "a conviction, strictly undemonstrable to science, that the universe has a direction and that it could—indeed if we are faithful it should—result in some sort of irreversible perfection." To Monod, all ideas of this kind are anathema; they are signs of "animism". Primitive man, as well as Leibnitz, Hegel, Spencer, Marx, Engels and Teilhard, are dumped on the animist heap. Not to be an animist you have to accept as an axiom that evolution has no direction. Nature is "objective", not "projective". Monod sees "pure chance, absolutely free but blind, at the very root of the stupendous edifice of evolution." Most important is that "man knows at last that he is alone in the universe's unfeeling immensity, out of which he emerged only by chance."

The problem of direction or "necessity" versus "chance" deserves most careful consideration. It can, I believe, be shown that this dichotomy, applied by Monod to biological as well as to human evolution, is spurious. Evolution is neither necessary, in the sense of being predestined, nor is it a matter of chance or accident. It is governed by natural selection, in which ingredients of chance and antichance are blended in a way which makes the dichotomy meaningless, and which renders evolution to be a creative process.

The universe is an evolving product of an evolutionary process. It is not an accident; it is an enterprise.

The directionality and progressiveness of general evolution are surely not just happy accidents. They stem from the nature of living matter and the laws of biology, first of all from natural selection. General evolution "had to" be progressive. This is just as remarkable, and if you wish mysterious, as, according to Einstein, the power of man-created mathematical concepts to describe natural phenomena. In this sense, what Monod calls "the ancient animist covenant between man and nature" is very much alive.

What's New in Science?

Teilhard, a paleontologist, and Monod, a molecular biologist, based their evolutionary extensions on the best science of their day (the 1930s and 1960s, respectively). The contrasting philosophies espoused by each are perennial. But what of the science?

The 1970s gave birth to two profound theories that inspired (and titled) the first anthology in this series, From Gaia to Selfish Genes: Selected Writings in the Life Sciences. *The Gaia theory of James Lovelock and the selfish gene theory of Richard Dawkins offer scientific grounding for a spectrum of worldviews. Cosmic optimists and cosmic pessimists and all manner of dispositions in between can find safe harbor in one or another extension of these ideas.*

Gaia theory calls on us to regard the sum of all life on earth (together with the atmosphere, oceans, and soils) as a living system in itself—not a cell, not an organism, but a living and breathing biosphere all the same. From a Gaian perspective, we humans live not so much on earth as within Gaia. The selfish gene theory, in contrast, urges us to regard ourselves as the servants of our genes. Bare-bones genes took to inventing bodies for offense and defense when the going got rough beyond the primordial soup. We are the product of an arms race, a never-ending struggle of genes working in shifting alliances to get hold of limited resources and propagate in limited space.

Before Gaia and selfish genes, biologists and others who chose to extend evolutionary theory into the realm of meaning were trapped in a species perspective. Evolution was viewed as the comings and goings of species and groups of species, like dinosaurs. By turning the focus of biology to the biosphere and the gene, Lovelock and Dawkins have given grounds for new cosmologies of life. Biosphere and gene pull our perspective in opposite directions, but both fundamentally get us out of our skins.

In the 1980s researchers with a shared interest in "complexity" began to have a presence in science. Using computers to construct populations of generic units interacting under a panoply of different rules, these cross-disciplinary scientists discovered the power of collective behavior, and they began to make concrete the magic of emergence.

Following are essays that take each of these advances—Gaia, selfish genes, and complexity—as a starting point for conjecturing how one might fashion a personal worldview that goes beyond the binary staked out by Pierre Teilhard de Chardin and Jacques Monod.

Untitled I (Marian
Vlasic)

Worldview Extensions of a Biospheric Perspective

Dorion Sagan, a science writer, has suggested that out of a biospheric perspective comes a veritable "ecological smorgasbord." In a contributed essay Sagan presents his personal favorites. But first, here is a mini-review, as I see it, of the feast of ideas we might choose from—the sweet, the sour, the salty.

Connie Barlow

A biospheric perspective can nurture a host of worldviews and value systems. One philosophical extension is decidedly misanthropic: humans are a blight, a cancer on the planet. Another is nihilistic: bacteria run the biosphere; the best or the worst we may do means nothing to these primal and ubiquitous cells, whose germ line will surely survive us for at least as long as it has preceded us. A biospheric perspective can also give the gift of community, nurturing a cross-species affection that knows no bounds. It can even conjure a stand-in mother, Gaia, eminently worthy of our love and respect. A deep-ecology version of a biospheric perspective—whether or not it acknowledges a global-scale Gaia—therefore offers one venue of scientific and emotional support for an ecologically attentive and caring outlook. Alarm for the health of the biosphere, coupled with a sadness for each species that now falls with the rain forests or the populations that fell with the forests and the open plains of Europe and North America, can be a source of meaning and a spur to commitment.

A biospheric perspective can also be given a slant that is not so far removed from humanist leanings. Vladimir Vernadsky, who pioneered the science of global biogeochemistry in Russia in the first half of this century, believed that humankind has a special role to play in the biosphere. By mining deposits of phosphates for fertilizer, irrigating arid lands, and putting vast quantities of iron and other minerals into partnership with protoplasm in the technosphere, humankind is increasing the amount of matter participating in biogeochemical cycles. Alarm about the amount and diversity of protoplasm we are now removing from those cycles, however, would temper current expression of a Vernadskian view.

A more expressly humanist view would regard our own species as a kind of sense organ or neural network of Gaia. James Lovelock has speculated, "Homo sapiens, with his technological inventiveness and his increasingly subtle communications network, has vastly increased Gaia's range of perception. She is now through us awake and aware of herself. She has seen the reflection of her fair face through the eyes of astronauts and the television cameras of orbiting spacecraft. Our sensations of wonder and pleasure, our capacity for conscious thought and speculation, our restless drive and curiosity are hers to share." Though grand, this Gaian perspective is more humble than the view of Julian Huxley and Teilhard de Chardin. Both felt that through humankind the *universe* was becoming conscious of itself.

Techno-anthropism is another possible extension of a biospheric perspective. Perhaps the role of humankind is to protect the biosphere from errant strikes of comets and giant meteors, the likes of which (according to David Raup and Michael Rampino) repeatedly pocked and pummeled earth in the past. As pluricellular creatures invented mineral shells for protection in the Cambrian, perhaps the biosphere is now at a similar cusp in its own evolution. If Michael Rampino is right about the misinterpretation of impact ejecta as glacial tillite, then—scandalous though the thought may be—perhaps humankind is an evolved adaptation of a biosphere that has been too successful in fixing into limestone and hydrocarbons the initial allotment of carbon dioxide that made the earth's primordial climate a hothouse and the Cretaceous a year-round sultry summer.

Finally, a kind of techno-cosmic extension of a biospheric perspective is feasible. In a 1990 book, *Biospheres: Reproducing Planet Earth,* Dorion Sagan suggests that it may be our fate to use our brains and our technology to get other biospheres going—spreading life, not necessarily intelligence, beyond our own world.

Here now is Dorion Sagan's take on a biospheric perspective.

Dorion Sagan

Vernadsky   Although coined by Austrian geologist Edward Seuss, the Russian crystallographer Vladimir Vernadsky brought the word biosphere into currency. Looking at life as a mineral phenomenon, Vernadsky did for space what Darwin did for time: he put life in a global context. Just as there was a lithosphere of rock and an atmosphere of air, so there was a biosphere of life—a planetary membrane not only where life exists, but which is itself life. It is remarkable that Vernadsky pictured life as a global phenomenon before any satellite photographs of Earth existed.

Teilhard probably took his conception of noosphere—a conscious layer of life—from Vernadsky, but he gave it a religious twist that suggested life is heading inevitably in the direction of cosmic perfection, the "omega point." The term noosphere was first used by Edouard Le Roy, philosopher Henri Bergson's successor at the Collège de France. Teilhard, Le Roy, and Vernadsky met in Paris for intellectual discussions in the 1920s but their use of the term noosphere, like their slants on evolution in general, differed. For Teilhard the noosphere was the "human" planetary layer forming "outside and above the biosphere," while for Vernadsky the noosphere referred to humanity and technology as an integral part of the planetary biosphere.

It is a shame that Teilhard's work is so much more well known than that of Vernadsky. As a thinker in a state (the former Soviet Union) that promoted atheism, Vernadsky developed a worldview free of organized religion's prejudices of teleology, that life has a purpose, or eschatology, that life is heading somewhere specific, for example, toward God.

Vernadsky portrayed life as a global phenomenon in which the sun's energy was transformed on Earth into a kind of "green fire." By this Vernadsky meant the photosynthetic growth of bacteria, algae, and plants. Vernadsky perceived the complex spread of disperse masses of organisms in the aggregate, as if ferns and foraminifera were merely particles in swirling films of matter. By stripping life of its vitalistic center, by looking at it as a mineral and examining its flows, Vernadsky deconstructed the lines between life and nonlife. This was a crucial step toward viewing life, as we often do today, in a global context. Lovelock's Gaia hypothesis also does this, but whereas Lovelock has likened the global environment to an organism, Vernadsky never spoke of life per se but only of "living matter" within the cycling biosphere. From different angles each view transcends both mechanism and animism.

Vernadsky perceived two trends in biospheric development, which he regarded as laws of nature. First, over time more and more kinds and quantities of elements are drawn into the biospheric flows. Second, the rates of the flows of these elements increase with time. For Vernadsky a flock of migrating gulls was a biospheric transport system for phosphorus (a limited element, used to make DNA, ATP, RNA, and other crucial biomolecules). Swarming locusts were not only the destroyers of croplands but geochemical agents carrying compounds in a direction opposite to that of gravity. The Vernadskian view therefore disrupts the boundaries of organic continuity simultaneously from below, by treating all organisms, including people, as collections of chemicals, and from above, by examining the global movements of living matter.

Narcissus   Teilhard's way of classifying living beings by their degree of cerebralization conforms to gut-level religious feelings that *Homo sapiens* cannot have arrived by chance. Monod, however, battles such notions with the claim that objectivity forces us to discard all such lingering vitalism and animism and consider ourselves as the outcome of chance and necessity. But peering at the blue surface of half the globe (not the "whole Earth" as is sometimes claimed), objectivity takes a backseat. We recognize that we are in the midst of something that transcends the scientific method of disconnected observation. Like Narcissus staring at his reflection in the water for the first time as Echo answers his calls from her hiding place behind the trees, we have glimpsed our full form in the mirror of satellite technology. In contrast to the cracked, cratery moon or the regolith of Mars, Earth (truly Ocean and Cloud) looks alive.

Just as we require Vernadsky to jettison the man-centeredness from the antiquated perspectives of Teilhard, so we must employ strong interpretations such as the ancient Greek myth of Narcissus to counter the naked existentialism of Monod. Freud used Oedipus, Narcissus, and other classical personalities to inquire into the functioning of the human psyche. The French pyschoanalyst Jacques Lacan posited a step in individual development he calls "the mirror stage." This stage is marked by the jubilation of the infant who, still lacking control of sense and limb, looks into the mirror (or at the mother's full body) and perceives a new wholeness. The new awe humanity takes in its global existence resembles such infantile joy. The dawning perception of the planet as a body is perhaps the greatest fruit of the space age. "We are living in a critical epoch of the history of mankind," wrote Vernadsky; what is emerging is "the idea and the feeling of the Whole."

Our opportunity to become Narcissus in space gives us grounds to transcend Monod. We do not have to construct meaning whole cloth out of pure will. Moreover, we do not have to turn inward to do it. We can be self-absorbed by turning outward if we expand our bounds of self. Monod, however, is right: we are partially creatures of chance. And yet, the free will we feel as artists, creators, interpreters, scientists, and, more generally, perceivers is no epiphenomenal fluke but a potential instrument of evolution. Our ability to choose and interpret can influence evolution in ways that go far beyond the bare landscape of a scientistic existentialism. This same perception that we inhabit a body made of all life also counters Teilhard's clumsy attempts to merge paleontology and Catholicism, by showing that our being is not central or divine, but integral and real.

Gaia   The expansion of self fostered by satellite perception is supported by science. The Gaia hypothesis holds that the surface of Earth has a physiology of its own. The chemistry of this global physiological system differs from a purely geological or geochemical system in that its chemical reactions are under active biological control. The Gaia hypothesis holds that aspects of the biosphere are regulated and modulated at a planetary level. Global temperature, the gaseous composition of the atmosphere, and the salinity and alkalinity of the oceans are all more than chance and necessity: they are all crucially influenced, and perhaps even orchestrated, by Earth's acting as a body. Evidence from such distinct disciplines as atmospheric chemistry, astronomy, and oceanography provides strong tentative support for the idea that the biosphere regulates these physical features, not necessarily consciously (as we are not necessarily conscious of our breathing or heart beating) but as the natural outcome of huge numbers of interacting, reproducing cells.

The Gaia concept resonates with ancient animistic beliefs. But far from resting on the mystic attributions of a new vitalism, Earth's "livingness" can be understood as the aggregate, emergent properties of the many gas-trading, gene-exchanging, growing organisms within it. From a planetary perspective, the phylogeny or evolution of species is akin to the development or ontogeny of Earth as a whole. We and our brethren species may or may not be loners among the stars, but there is no doubt that we are functionaries within a global body with its own complex physiological wisdom. This shift in perspective has the potential to change every domain of human endeavor. Science, for example, comes to resemble not the highest intellectual activity of a species isolated in its intelligence but rather the emergent perceptual and thinking processes of a growing and technological superorganism.

The Birth of New Biospheres   The biospheric perspective admits human embeddedness in a global living system in a way that makes us neither so special as Teilhard portrays nor so arbitrary as Monod envisions. We recognize, rather, that our perception of our place and role influences the reality of our place and role. If we choose to view ourselves as stewards coauthoring evolution through a manipulation of genes, that perspective will come into general currency and perhaps become the social reality through which we see ourselves. If we choose to view ourselves as egocentric adjuncts thrown off by an arbitrary, mechanical process of chance and necessity, then that will enter into our fate, at least in terms of self-perception.

We as a culture are in a period of massive shiftings and blendings of worldviews akin to the syncretism and blossoming of beliefs that occurred at the end of classical Greece and the fall of Rome. This rampant sprouting of belief systems during periods of social change resembles the acceleration in mutations and production of DNA fragments by bacteria exposed to strong ultraviolet radiation. Perhaps both are "brainstorming" operations for offering up more (and admittedly random) possibilities during times of duress.

Untitled II (Marian
Vlasic)

As we enter into a phase of globally connected human cultures, our relations with other organisms have come under scrutiny. A new concern for our relationships with other animals, plants, microbes, and the environment at large seems a natural outcome—the belated realization of a species that has spread in short order from several thousands to several billions, and with a presence now felt planetwide. From a planetary perspective, one of the most significant developments in our time is the production by NASA, the space industries of other countries, and even by private concerns of closed ecological systems. Communities of organisms are selected for their ability to effect complete cycles of biologically important compounds in containers closed to all matter exchange but open to sunlight. The simplest versions are sealed test tubes containing only ecosystems of bacteria, Earth's most metabolically diverse life forms. Ecospheres, glass paperweights containing a few shrimp and algae as well as bacteria, are popular novelty items. The biggest forms of closed ecological systems are being designed to contain people. To house human beings in space, however, these ecospheres will have to become miniature biospheres. Sometimes called "artificial" to distinguish them from the natural biosphere of the original, global ecosystem, these designed biospheres must be large and diverse. In addition to bacteria, fungi, protists, plants, and animals, biospheres capable of hosting humans will contain computer and communications systems, a little noosphere of their own.

To my mind the existence of a self-supporting ecosystem floating in space or settled comfortably on the surface of another world would represent nothing less than the reproduction of Gaia. Indeed, this thought experiment helps support the Gaia hypothesis, as the argument has been raised that Earth cannot be alive like an organism if it is incapable of reproducing. A reproducing Gaia should help dispel our fear of being galactic loners in sole possession of the enigmatic quality we associate with the cerebrum and call intelligence. Part of the wisdom of the body is to expand its organization, to perpetuate itself by budding and reproduction. The minimal unit imaginable for life to survive beyond Earth merges bacteria, fungi, plants, and humans into self-supporting communities. In the evolutionary long run, humanity itself will survive only as integral parts of a wild nexus of widely divergent life forms whose reproduction becomes possible only in concert.

———

The vast Pre-Cambrian microbial mats,
Exiled, enfossiled into stromatolites,
Barely survived four million millennia
Spread thin through soil and water.

Yet their necessity still haunts our intestines,
Bonding the vegetable in matrimony

To that noble gas nitrogen.
They practice a most sophisticated chemistry.

Do not discount their gypsy influence.
Bacteria may have engineered us solely
To explore lunar dust and Martian clays
For their most distant of cousins.

—*Robert Frazier*

———

Worldview Extensions of a Genic Perspective

Richard Dawkins, in his 1976 book The Selfish Gene, *beckons us to craft cosmologies from new knowledge—and a wholly new perspective—of life at the smallest scale. Here too, one can partake of a smorgasbord of worldview extensions, as I outline next.*

Connie Barlow

At its bleakest, the selfish gene theory promotes a worldview that finds no meaning in anything above the level of genes. Helices of DNA grouped into genes and bound into chromosomes do not, in this view, exist so that individual humans can face death consoled that their progeny live on. Genes do not exist to ensure a steady supply of proud grandparents. They do not even exist in order to keep the species going. Rather, we exist for them.

We and all other organisms are, in the words of Richard Dawkins, mere "survival machines—robot vehicles blindly programmed to preserve the selfish molecules known as genes." Those selfish molecules built us only because of the press of arms races. The first arms race flared, in Dawkins's view, even before nucleic acids and proteins produced the first cell. Eventually the message would come: build a body or die. Worse, build a body or fail to reproduce—which is worse because, in this perspective, reproduction is the very essence of life.

It was the drive to reproduce, after all, that gave any meaning to the idea of scarcity. And scarcity promoted innovation. And it was innovation that made things interesting. Reproduction, fecundity kept feeding the fire. It gave natural selection something to select among. Its triumph: some four billion years of uninterrupted genic continuity.

So genes built bodies, and they came together in symbiotic legions in order to do so. We are "their" bodies far more than they are "our" genes. Our role is to propagate the genes within us—or the genes within another body that manage to hijack us for their particular use. Richard Dawkins illustrates such cross-species genic manipulation with the example of a tiny parasite that finds sustenance alternately within the bodies of ants and sheep. A parasitized ant, unlike its busy comrades, is induced by the parasite that bores into its brain to perch on the tip of a blade of grass and simply pass the hours there—in the best possible spot to be consumed by the parasite's next quarry. A familiar form of cross-body genic manipulation is what we humans call romance.

The selfish gene theory is not, however, strictly the province of cosmic pessimists and misanthropes. Another legitimate extension sanctions our deepest and most traditional beliefs and drives. Anyone who finds fulfillment in parenthood knows the meaning of life. Prince Charles may have his difficulties, but he has his heirs.

We can also put our energies into propagating our beloved "memes." Richard Dawkins invented this delightful and useful word to signify the cultural artifacts that have transcended the evolutionary stream of "wet" biology. Memes now partake of a kind of evolution of their own. They can be as laudatory as a painting in the Louvre, the idea of democracy, the film *E.T.* They can also be as irritating as elevator music and the shoelaces a fashion-conscious six-year-old refuses to tie. As with genes, successful memes are those that have an ability to propagate. And clonal memes, like genes, can find an immortality that complex and vulnerable bodies are denied. What is more, they surpass genes in their ability to reproduce fast, notably by way of electronics.

I am Napir-Asu, wife of Untash Napirisha. He who would seize my statue, who would smash it, who would destroy its inscription, who would erase my name, may be smitten by the curse of Napirisha, of Kiririsha, and of Inshushinak, that his name shall become extinct, that his offspring be barren.

This inscription is engraved on the skirt of a life-size bronze statue of a Mesopotamian queen who lived 3,400 years ago. That statue has indeed been smashed (it is missing head and arm) and it has been seized (by the Louvre), but it is testament to the hold that genes and memes claim on the human psyche.

Whether one subscribes to the punishing worldview of Napir-Asu, to the apple pie sanctity of family values, or to the memic obsessions of artists, writers, and scientists, the selfish gene (and meme) theory tells us that what we are doing is normal and, in an evolutionary sense, fitting and right. The flaunted or veiled Me, Me, Me of those drives can also be paired with a worldview derived from a biospheric perspective. Perhaps a sense of righteousness for rebelling against our lords and holding one's progeny to two (or none). Perhaps a mission to heal what we have harmed, each in our own small way. Or perhaps a reveling in a reborn sense of community with our biospheric and genic siblings.

Worldview Extensions of Complexity Theory

In 1992 a triad of new books made a kind of coming-out party for the cross-disciplinary science of complexity. M. Mitchell Waldrop, science writer and author of Complexity: The Emerging Science at the Edge of Order and Chaos, *joins with me in the following essay to present the essence of this research focus and what it may mean for those of us who choose to craft our worldviews, at least in part, from the science of our day.*

Connie Barlow, M. Mitchell Waldrop

Jacques Monod claimed that "from a source of noise natural selection alone and unaided [has] drawn all the music of the biosphere." Scientists and philosophers who disagree have, from Monod's viewpoint, "abandoned the postulate of objectivity." They have abandoned science.

Nonetheless, in every generation since Darwin some biologists and philosophers have judged that, while chance variation and natural selection are all good and well, something more, something perhaps deeper assists in the genesis of complexity. Some dissenters believed they had, in fact, glimpsed the "something more." The protest of others was more a visceral response to what was perceived as a narrow-minded view of causation: the hubris of finding a single cause and then finding it satisfactory.

But if there is something more, what is it? Bergson's claim for an "élan vital" and Teilhard's claim for a "radial energy" fell at the fringes of science because a name is not an explanation. Do today's critics of chance and necessity have any better grounding in objective fact?

Maybe.

In the 1980s researchers from a range of academic disciplines began to bring together a mixture of old and new ideas to improve the framework for understanding complex systems—not just the cells and organisms and ecosystems of the biological sciences but the complex systems of societies, economies, and even minds.

From biology came the insight that had long ago transformed natural history into the explanatory science of evolutionary biology: natural selection. In its early guise natural selection was the story that did in divine design. Today, the compass of natural selection reaches beyond organic adaptation to include "learning" in a host of contexts. From biology, too, comes the insight of population thinking. In the first half of this century biologists were able to forge a synthesis of Darwinian natural selection with Mendelian genetics by working through the mathematics of genetic evolution in an interbreeding population rather than in a single lineage. With the aid of personal computers, complexity researchers now are able to explore population dynamics at a level inaccessible to mathematical solution.

The computer age has provided a means not only for exploring but for expressing questions in complexity research. Advances in the computational sciences include an analytical mindset and a rigorous language for describing processes that unfold through time—sequences of decisions and sequences of actions. The ability to present the ongoing computations onscreen gives us a sixth sense, a way to use the superb visual powers granted us by evolution to perceive, in a gestalt way, the flux of pattern, the behavioral richness that is the very essence of complexity.

"Gliders," for example, wing their way across the screen of the Game of Life, while "glider guns" maintain the supply. In another artificial life program a flock of "boids" negotiates obstacles and maintains cohesiveness; the flocking behavior emerges out of just a three-part rule that each boid follows in reacting to decisions made by its immediate neighbors.

Gliders and boids show that simple rules can lead to astonishing behavior. A population of singularly stupid and utterly provincial agents following mind-numbingly simple and mechanical rules can give rise to collective behavior that is surprising in its richness, in its lifelike qualities. Adam Smith recognized the power of this kind of bottom-up emergence when he spoke of the Invisible Hand. And not for nothing have naturalists long marveled at the so-called balance of nature.

With computers one thus can chart the patterns of subscale change that sweep through a population when components act and interact not sequentially but in parallel. One does not, moreover, corner complexity in some sort of end state. It cannot be teased over to one side of an equal sign. Complexity is not a thing; it is a behavior. It is as much an expression in time as is music or conversation. Nevertheless, complexity researchers owe to applied mathematics an understanding of nonlinear dynamical systems—systems in which the whole is quite literally greater than the sum of its parts. Chaos and catastrophe theory are perhaps the best-known fruits of this research.

Finally, out of the physical sciences come nonequilibrium thermodynamics and chemical self-catalysis. Dissipative structures (including whirlpools and thermal convection cells) and autocatalytic chemical structures (for example, molecular hypercycles) are seductively lifelike in their ability to arise and persist in a surround of environmental noise. Whirlpools and self-catalyzing loops of chemicals are not so much selected by the environment as bootstrapped into existence. Whether the root cause of this self-organization is some sort of internal selection (selection going on at a lower level) or something altogether different, complexity researchers tend to agree that these classes of complex systems are expressive of something more than the machinations of environmental chance and necessity.

Attempts to use these ideas and tools to break through to new levels of understanding have only partly borne fruit—as yet. Nonetheless, enough is known, enough is suggestive to give a hint that Monod and Teilhard may both be right and that Dawkins and Lovelock, too, may have a finger on the pulse of the universe. For example, Monod is quite correct in seeing aimless chance and plodding necessity as the matrix out of which comes everything interesting and lovable in the universe. But within that matrix can be found pockets of local activity where thermodynamic, chemical, and (well into the tree of life) even willful behavior seem to be orchestrating the rise of complexity—the *spontaneous* rise of complexity.

Within the life sciences, proponents of self-organization have their best chances to be heard at the lower and upper reaches of complex living systems. Speciation was no mystery for Darwin, once he had natural selection in hand. But Darwin never was sure that his great insight had much to say about the origin of life. Even today there exists a vast chasm between evolutionary biologists and origins-of-life researchers. But a growing knowledge of thermodynamic and molecular forms of self-organization may eventually carry physicochemical theory up to the very threshold of life. Meanwhile, a population approach for the rise of more and more lifelike behavior may also help connect the rich chemistry of the primordial soup with the beginnings of biology. A populational model has, for example, recently become widely accepted as the explanation for the intelligence of our own immune systems. More controversially, such a model may explain our minds.

At the upper reaches of life, ecosystems, complexity researchers are also making headway. Ecologist Tom Ray got the shock of his life when "Tierra," his first attempt to grow an ecosystem virtually *ex nihilo* in his computer became an echo of the real world. The initial 80-command "organism," whose only goal in life is to replicate, quickly speciated into bigger and smaller lengths of commands, each finding ways (some quite ingenious in their efficiency) to reproduce. These replicators in their digital brew competed for the limited resource of computational time in which to execute their chains of commands. The scythe of a grim reaper at the end of the queue also pressed upon them.

The speciation in Ray's Tierra was fascinating, but the ecological results were astounding. Parasites arose quickly. Then the teaming of symbionts proved a way to ward them off—for a while. Parasites of parasites arose, and parasites of parasites of parasites, and even symbioses of parasites. Extinctions pruned the diversity and opened niches; new species, symbionts, and parasites emerged. What is more, the ecosystem never settled down. It was a world of perpetual novelty, an endless shuffling of identities and relationships, a tapestry woven on the warp and woof of strife and synergy.

Tom Ray's artificial ecosystem was as close to a display of "emergence" as this kind of computer operation has ever issued up. Yet it was governed only by the necessity of simple rules that quickly led to a population of interacting agents. It also owed to chance—to a programmed level of randomizing events to ensure recombinations and mutations of commands. And all this fed off the particular happenstance of the ancestral replicator. But trial after trial, beginning with different ancestors, the same ways of making a living emerged. Never repetitive, there was nevertheless a pattern of relationships that shuffled and reappeared in different guises. The artificial world of Tierra in this way resembled Teilhard's vision of a preferred, even predestined, pathway. Ray's experiment hints that the ways creatures make a living on Earth may not be just an expression of the particular chance and necessity of complex chains of hydrocarbons interacting on the surface of the third planet from a particular Main Sequence star on the outskirts of a certain spiral galaxy.

Tierra lends insight to the workings of ecosystems, but what of the biggest living system of all, the biosphere? What can the complexity sciences tell us about it? Perhaps this kind of research can provide the criteria for distinguishing whether the biosphere is indeed an "it" or an aggregate. Are there global patterns in space and time that show the mark of a complex adaptive system? If so, how does the biosphere, absent an external selective force, organize itself into a coherent system?

The complexity sciences today thus offer opportunities for discovery like that of the distant continents in the days of the HMS *Beagle*. South America, Malaysia, and other exotic lands offered the abundance and novelty for European botanists and zoologists to engage in their descriptive tasks, their cataloging of the biosphere's diversity. Complex systems are still rich for description and for moving into a more fundamental level of description focusing on the conditions that give rise to such systems, which in turn might secure wider recognition of this budding research program. For example, an idea that is sometimes called "edge of chaos," sometimes "self-organized criticality," might take complexity experimentation into the realm of explanatory science. Chris Langton, a computer scientist, is one of the key contributors to the "edge of chaos" idea. Langton discovered that complex adaptive systems—those populations of simple interacting units that respond globally to environmental perturbances by changing, by adapting, and that trigger their own transformations—are found in a precisely identifiable computational region between the too ordered and the too anarchic.

Complex adaptive systems turn out to be one of four types of universal behaviors in computer worlds composed of populations of interacting units. If the network of relationships has little flexibility, if it is highly ordered, a point attractor is the result. When this kind of system is shown onscreen, a single state emerges and then freezes; nothing else happens from then on. A little loosening of system parameters gives rise to periodic attractors, a sequence of different states that unfolds in an unchanging, repetitive order. On the other hand, too much freedom in the relationships of interacting parts gives rise to strange attractors, also known as chaos. Displayed graphically onscreen, a strange attractor is obviously repetitive in broad terms, but each expression is in some way unique. What is more, the next expression cannot be predicted mathematically; it can be found only by running the program.

Complex adaptive systems lie in the computational region between the too-ordered state of point and periodic attractors and the too-anarchic state of chaotic attractors. In Langton's terminology complex adaptive systems reside "at the edge of chaos." The transition from order to complexity, or from chaos to complexity, is not gradual. It is sudden; graphically portrayed, it is nonlinear. At these boundaries a kind of "phase transition" occurs. Complex adaptive systems are not just a blend of order and chaos; they are order and chaos in a very precise kind of dynamic tension.

Physicist Per Bak's idea of "self-organized criticality" is similar. In his work he uses a sandpile as both model and metaphor. As grain after grain rains down from above, the pile grows higher and wider, all the while keeping its conical shape. If the pile is built on a tabletop, an additional grain eventually will unleash a cascade that sends not one grain but an avalanche cascading off the edge. More sand grains can then accumulate to no effect before one will at last trigger another, bigger or smaller, avalanche. Extended into human terms, self-organized criticality means that any one of us—however humble our actions and aspirations—could be in a position to induce big changes, for better or for worse.

But agents need not be just faceless and expendable clones. Organisms are not passive grains of sand. As Alister Hardy and Karl Popper explained in chapter 5, initiative can play a powerful role in evolution. And by way of culture human memes can go far beyond the constrictions of our own time and place. That nubbin of an idea we send off into the world today may fall to no effect on the cultural heap, but it might be picked up by somebody else—perhaps decades later—and launched again to much greater effect.

Strife, synergy, and initiative thus all seem to play a role in the genesis of complexity. And within the broad framework of complex adaptive systems, one can see value in the ideas and extensions of both Teilhard and Monod, of Lovelock and Dawkins.

Teilhard had a masterly grasp of the power of collective action and the role of emergence in natural systems. The coherence he identified in the biosphere and the noosphere seem to be rooted in reality. One need not believe in Omega in order to extract value from the Teilhardian vision. At the same time, the aimless chance and the plodding necessity portrayed by Jacques Monod are, at base, what the universe is built from. And objectivity still reigns in science.

Lovelock's Gaia, if it/she exists, would be the biggest complex adaptive system we know. The scientific controversy about Gaia, moreover, is an obvious place to apply the fruits of complex systems theory—to search for expression in an actual system the causes that show so clearly in artificial systems. Meanwhile, Dawkins's insistence on the power of replication seems to have been vindicated in Tom Ray's world of Tierra. And his portrayal of the power and emergence of memes, of the rightness of regarding them as new units of evolution, has gained acceptance in just about all camps.

There may be lots of ways, in biological fact and philosophical extension, to transcend the binary of Teilhard and Monod, of Lovelock and Dawkins. But the study of complex systems seems to provide an especially promising path.

———

The gold-dust comes to birth
with the quartz-sand all around it,
and this is as much a condition of religion
as of any other excellent possession.
There must be extrication;
there must be competition for survival;
but the clay matrix and the noble gem
must first come into being unsifted.

—*William James*

IV Evolution and Religion

The evolutionary epic is probably the best myth we will ever have.

—*Edward O. Wilson, 1978*

Evolution is surely a religion, in every sense of the word. It is a world view, a philosophy of life and meaning, an attempt to explain the origin and development of everything, from elements to galaxies to people, without the necessity of an omnipotent, personal, transcendent Creator. It is absurd for evolutionists to insist, as they often do, that evolution is science and creation is religious.

—*Henry M. Morris, 1985*

We are in the midst of a revelatory experience of the universe that must be compared in its magnitude with those of the great religious revelations. And we need only wander about telling this new story to ignite a transformation of humanity.

—*Brian Swimme, 1988*

"The evolutionary epic is probably the best myth we will ever have." So proclaims biologist Edward O. Wilson in his Pulitzer Prize–winning book, On Human Nature. *In one chapter Wilson deploys sociobiology (a field he founded) to dissect religious myths and practices for their adaptive value. Excerpts from* On Human Nature *(copyright 1978 by the President and Fellows of Harvard College) are reprinted by permission of the author and Harvard University Press, Cambridge, Massachusetts.*

Inadequacies of Humanism and Process Theology

Edward O. Wilson   The predisposition to religious belief is the most complex and powerful force in the human mind and in all probability an ineradicable part of human nature. Emile Durkheim, an agnostic, characterized religious practice as the consecration of the group and the core of society. It is one of the universals of social behavior, taking recognizable form in every society from hunter-gatherer bands to socialist republics. Its rudiments go back at least to the bone altars and funerary rites of Neanderthal man. At Shanidar, Iraq, sixty thousand years ago, Neanderthal people decorated a grave with seven species of flowers having medicinal and economic value, perhaps to honor a shaman. Since that time, according to the anthropologist Anthony F. C. Wallace, mankind has produced on the order of one hundred thousand religions.

Skeptics continue to nourish the belief that science and learning will banish religion, which they consider to be no more than a tissue of illusions. The noblest among them are sure that humanity migrates toward knowledge by logotaxis, an automatic orientation toward information, so that organized religion must continue its retreat as darkness before enlightenment's brightening dawn.

In his *System of Positive Polity*, published between 1846 and 1854, Auguste Comte argued that religious superstition can be defeated at its source. He recommended that educated people fabricate a secular religion consisting of hierarchies, liturgy, canons, and sacraments not unlike those of Roman Catholicism, but with society replacing God as the Grand Being to worship. Today, scientists and other scholars, organized into learned groups such as the American Humanist Society and Institute on Religion in an Age of Science, support little magazines distributed by subscription and organize campaigns to discredit Christian fundamentalism, astrology, and Immanuel Velikovsky. Their crisply logical salvos, endorsed by whole arrogances of Nobel Laureates, pass like steel-jacketed bullets through fog. The humanists are vastly outnumbered by true believers, by the people who follow Jeane Dixon but have never heard of Ralph Wendell Burhoe. Men, it appears, would rather believe than know. They would rather have the void as purpose, as Nietzsche despairingly wrote so long ago when science was at its full promise, than be void of purpose.

Other well-meaning scholars have tried to reconcile science and religion by compartmentalizing the two rivals. Newton saw himself not only as a scientist but as a historical scholar whose duty was to decipher the Scriptures as a true historical record. Although his own mighty effort created the first modern synthesis of the physical sciences, he regarded that achievement as only a way station to an understanding of the supernatural. The Creator, he believed, has given the scholar two works to read, the book of nature and the book of scriptures. Today, thanks to the relentless advance of the science which Newton pioneered, God's immanence has been pushed to somewhere below the subatomic particles or beyond the farthest visible galaxy.

This apparent exclusion has spurred still other philosophers and scientists to create "process theology," in which God's presence is inferred from the inherent properties of atomic structure. As conceived originally by Alfred North Whitehead, God is not to be viewed as an extraneous force, who creates miracles and presides over the metaphysical verities. He is present continuously and ubiquitously. He covertly guides the emergence of molecules from atoms, living organisms from molecules, and mind from matter. The properties of the electron cannot be finally announced until their end product, the mind, is understood. Process is reality, reality process, and the hand of God is manifest in the laws of science. Hence religious and scientific pursuits are intrinsically compatible, so that well-meaning scientists can return to their calling in a state of mental peace. But all this, the reader will immediately recognize, is a world apart from the real religion of the aboriginal corroboree and the Council of Trent.

The Sociobiology of Religion

Today, as always before, the mind cannot comprehend the meaning of the collision between irresistible scientific materialism and immovable religious faith. We try to cope through a step-by-step pragmatism. Our schizophrenic societies progress by knowledge but survive on inspiration derived from the very beliefs which that knowledge erodes. I suggest that the paradox can be at least intellectually resolved, not all at once but eventually and with consequences difficult to predict, if we pay due attention to the sociobiology of religion. Although the manifestations of the religious experience are resplendent and multidimensional, and so complicated that the finest of psychoanalysts and philosophers get lost in their labyrinth, I believe that religious practices can be mapped onto the two dimensions of genetic advantage and evolutionary change. Let me moderate this statement at once by conceding that if the principles of evolutionary theory do indeed contain theology's Rosetta stone, the translation cannot be expected to encompass in detail all religious phenomena. By traditional methods of reduction and analysis science can explain religion but cannot diminish the importance of its substance.

The deep structure of religious belief can be probed by examining natural selection at three successive levels. At the surface, selection is ecclesiastic: rituals and conventions are chosen by religious leaders for their emotional impact under contemporary social conditions. Ecclesiastic selection can be either dogmatic and stabilizing or evangelistic and dynamic. In either case the results are culturally transmitted; hence variations in religious practice from one society to the next are based on learning and not on genes.

At the next level selection is ecological. Whatever the fidelity of ecclesiastic selection to the emotions of the faithful, however easily its favored conventions are learned, the resulting practice must eventually be tested by the demands of the environment. If religions weaken their societies during warfare, encourage the destruction of the environment, shorten lives, or interfere with procreation they will, regardless of their short-term emotional benefits, initiate their own decline.

Finally, in the midst of these complicated epicycles of cultural evolution and population fluctuation, the frequencies of genes are changing. The hypothesis before us is that some gene frequencies are changed in consistent ways by ecclesiastic selection. Human genes, it will be recalled, program the functioning of the nervous, sensory, and hormonal systems of the body, and thereby almost certainly influence the learning process. They constrain the maturation of some behaviors and the learning rules of other behaviors. Incest taboos, taboos in general, xenophobia, the dichotomization of objects into the sacred and profane, nosism, hierarchical dominance systems, intense attention toward leaders, charisma, trophyism, and trance-induction are among the elements of religious behavior most likely to be shaped by developmental programs and learning rules. All of these processes act to circumscribe a social group and bind its members together in unquestioning allegiance.

While growing increasingly sophisticated, anthropology and history continue to support Max Weber's conclusion that the more elementary religions seek the supernatural for purely mundane rewards: long life, abundant land and food, averting physical catastrophes, and the conquest of enemies. The highest forms of religious practice can be seen to confer biological advantage. Above all they congeal identity. In the midst of the chaotic and potentially disorienting experiences each person undergoes daily, religion classifies him, provides him with unquestioned membership in a group claiming great powers, and by this means gives him a driving purpose in life compatible with his self-interest. His strength is the strength of the group, his guide the sacred covenant. The theologian and sociobiologist Hans J. Mol has aptly termed this key process the "sacralization of identity."

A kind of cultural Darwinism also operates during the competition among sects in the evolution of more advanced religions. Those that gain adherents grow; those that cannot, disappear. Consequently religions are like other human institutions in that they evolve in directions that enhance the welfare of the practitioners. Because this demographic benefit must accrue to the group as a whole, it can be gained partly by altruism and partly by exploitation, with certain sectors profiting at the expense of others. Alternatively, the benefit can arise as the sum of the generally increased fitnesses of all of the members. The resulting distinction in social terms is between the more oppressive and the more beneficent religions. All religions are probably oppressive to some degree, especially when they are promoted by chiefdoms and states.

There is a principle in ecology, Gause's law, which states that maximum competition is to be found between those species with identical needs. In a similar manner, the one form of altruism that religions seldom display is tolerance of other religions. Their hostility intensifies when societies clash, because religion is superbly serviceable to the purposes of warfare and economic exploitation. The conqueror's religion becomes a sword, that of the conquered a shield.

The Mythopoeic Drive

As science proceeds to dismantle the ancient mythic stories one by one, theology retreats to the final redoubt from which it can never be driven. This is the idea of God in the creation myth: God as will, the cause of existence, and the agent who generated all of the universe in the original fireball and set the natural laws by which the universe evolved. So long as the redoubt exists, theology can slip out through its portals and make occasional sallies back into the real world. Whenever other philosophers let their guard down, deists can, in the manner of process theology, postulate a pervasive transcendental will. They can even hypothesize miracles.

God remains a viable hypothesis as the prime mover, however undefinable and untestable that conception may be. The rituals of religion, especially the rites of passage and the sanctification of nationhood, are deeply entrenched and incorporate some of the most magnificent elements of existing cultures. They will certainly continue to be practiced long after their etiology has been disclosed. The anguish of death alone will be

enough to keep them alive. It would be arrogant to suggest that a belief in a personal, moral God will disappear, just as it would be reckless to predict the forms that ritual will take as scientific materialism appropriates the mythopoeic energies to its own ends.

I am not suggesting that scientific naturalism be used as an alternative form of organized formal religion. My own reasoning follows in a direct line from the humanism of the Huxleys, Waddington, Monod, Pauli, Dobzhansky, Cattell, and others who have risked looking this Gorgon in the face. Each has achieved less than his purpose, I believe, for one or the other of two reasons. He has either rejected religious belief as animism or else recommended that it be sequestered in some gentle preserve of the mind where it can live out its culture-spawned existence apart from the mainstream of intellectual endeavor.

Humanists show a touching faith in the power of knowledge and the idea of evolutionary progress over the minds of men. I am suggesting a modification of scientific humanism through the recognition that the mental processes of religious belief—consecration of personal and group identity, attention to charismatic leaders, mythopoeism, and others—represent programmed predispositions whose self-sufficient components were incorporated into the neural apparatus of the brain by thousands of generations of genetic evolution. As such they are powerful, ineradicable, and at the center of human social existence. They are also structured to a degree not previously appreciated by most philosophers. I suggest further that scientific materialism must accommodate them on two levels: as a scientific puzzle of great complexity and interest, and as a source of energies that can be shifted in new directions when scientific materialism itself is accepted as the more powerful mythology.

When blind ideologies and religious beliefs are stripped away, others are quickly manufactured as replacements. If the cerebral cortex is rigidly trained in the techniques of critical analysis and packed with tested information, it will reorder all that into some form of morality, religion, and mythology. If the mind is instructed that its pararational activity cannot be combined with the rational, it will divide itself into two compartments so that both activities can continue to flourish side by side.

This mythopoeic drive can be harnessed to learning and the rational search for human progress if we finally concede that scientific materialism is itself a mythology defined in the noble sense. So let me give again the reasons why I consider the scientific ethos superior to religion: its repeated triumphs in explaining and controlling the physical world; its self-correcting nature open to all competent to devise and conduct the tests; its readiness to examine all subjects sacred and profane; and now the possibility of explaining traditional religion by the mechanistic models of evolutionary biology. The last achievement will be crucial. If religion, including dogmatic secular ideologies, can be systematically analyzed and explained as a product of the brain's evolution, its power as an external source of morality will be gone forever.

Make no mistake about the power of scientific materialism. It presents the human mind with an alternative mythology that until now has always, point for point in zones of conflict, defeated traditional religion. Its narrative form is the epic: the evolution of the universe from the big bang of fifteen billion years ago through the origin of the elements and celestial bodies to the beginnings of life on earth.

The evolutionary epic is mythology in the sense that the laws it adduces here and now are believed but can never be definitely proved to form a cause-and-effect continuum from physics to the social sciences, from this world to all other worlds in the visible universe, and backward through time to the beginning of the universe. Every part of existence is considered to be obedient to physical laws requiring no external control. The scientist's devotion to parsimony in explanation excludes the divine spirit and other extraneous agents. Most importantly, we have come to the crucial stage in the history of biology when religion itself is subject to the explanations of the natural sciences. Sociobiology can account for the very origin of mythology by the principle of natural selection acting on the genetically evolving material structure of the human brain.

If this interpretation is correct, the final decisive edge enjoyed by scientific naturalism will come from its capacity to explain traditional religion, its chief competitor, as a wholly material phenomenon. Theology is not likely to survive as an independent intellectual discipline. But religion itself will endure for a long time as a vital force in society. Like the mythical giant Antaeus who drew energy from his mother, the earth, religion cannot be defeated by those who merely cast it down.

Promoting the Evolutionary Epic

The spiritual weakness of scientific naturalism is due to the fact that it has no primal source of power. While explaining the biological sources of religious emotional strength, it is unable in its present form to draw on them, because the evolutionary epic denies immortality to the individual and divine privilege to the society, and it suggests only an existential meaning for the human species. Humanists will never enjoy the hot pleasures of spiritual conversion and self-surrender; scientists cannot in all honesty serve as priests. So the time has come to ask: Does a way exist to divert the power of religion into the services of the great new enterprise that lays bare the sources of that power?

What I am suggesting, in the end, is that the evolutionary epic is probably the best myth we will ever have. It can be adjusted until it comes as close to truth as the human mind is constructed to judge the truth. And if that is the case, the mythopoeic requirements of the mind must somehow be met by scientific materialism so as to reinvest our superb energies.

There are ways of managing such a shift honestly and without dogma. One is to cultivate more intensely the relationship between the sciences and humanities. The great British biologist J. B. S. Haldane said of science and literature, "I am absolutely convinced that science is vastly more stimulating to the imagination than are the classics, but the products of the stimulus do not normally see the light because scientific men as a class are devoid of any perception of literary form."

Indeed, the origin of the universe in the big bang of fifteen billion years ago, as deduced by astronomers and physicists, is far more awesome than the first chapter of Genesis or the Ninevite epic of Gilgamesh. When the scientists project physical processes backward to that moment with the aid of mathematical models they are talking about everything—literally everything—and when they move forward in time to pulsars, supernovas, and the collision of black holes they probe distances and mysteries beyond the imaginings of earlier generations. Recall how God lashed Job with concepts meant to overwhelm the human mind:

Who is this whose ignorant words cloud my design in darkness?
Brace yourself and stand up like a man;
I will ask questions, and you shall answer . . .
Have you descended to the springs of the sea or walked in the unfathomable deep?
Have the gates of death been revealed to you?
Have you ever seen the door-keepers of the place of darkness?
Have you comprehended the vast expanse of the world?
Come, tell me all this, if you know.

And yes, we do know and we have told. Jehovah's challenges have been met and scientists have pressed on to uncover and to solve even greater puzzles. The physical basis of life is known; we understand approximately how and when it started on earth. New species have been created in the laboratory and evolution has been traced at the molecular level. Genes can be spliced from one kind of organism into another. Molecular biologists have most of the knowledge needed to create elementary forms of life. Our machines, settled on Mars, have transmitted panoramic views and the results of chemical soil analysis. Could the Old Testament writers have conceived of such activity? And still the process of great scientific discovery gathers momentum.

Unknown and surprising things await. They are as accessible as in those days of primitive wonder when the early European explorers went forth and came upon new worlds and the first microscopists watched bacteria swim across drops of water. As knowledge grows science must increasingly become the stimulus to imagination.

In the spirit of the enrichment of the evolutionary epic, modern writers often summon the classical mythic heroes to illustrate their view of the predicament of humankind: the existential Sisyphus, turning fate into the only means of expression open to him; hesitant Arjuna at war with his conscience on the Field of Righteousness; disastrous Pandora bestowing the ills of mortal existence on human beings; and uncomplaining Atlas, steward of the finite Earth. Prometheus has gone somewhat out of fashion in recent years as a concession to resource limitation and managerial prudence. But we should not lose faith in him. Come back with me for a moment to the original, Aeschylean Prometheus:

Chorus: Did you perhaps go further than you have told us?
Prometheus: I caused mortals to cease foreseeing doom.
Chorus: What cure did you provide them with against that sickness?
Prometheus: I placed in them blind hopes.

The true Promethean spirit of science means to liberate man by giving him knowledge and some measure of dominion over the physical environment. But at another level, and in a new age, it also constructs the mythology of scientific materialism, guided by the corrective devices of the scientific method, addressed with precise and deliberately affective appeal to the deepest needs of human nature, and kept strong by the blind hopes that the journey on which we are now embarked will be farther and better than the one just completed.

———

The world is too much with us; late and soon,
Getting and spending, we lay waste our powers.
Little we see in Nature that is ours;
We have given our hearts away, a sordid boon!
This Sea that bares her bosom to the moon;
The winds that will be howling all hours
And are up-gathered now like sleeping flowers;
For this, for everything, we are out of tune;
It moves us not.—Great God! I'd rather be
A pagan suckled in a creed outworn;
So might I, standing on this pleasant lea,
Have glimpses that would make one less forlorn;
Have sight of Proteus rising from the sea;
Or hear old Triton blow his wreathed horn.

—*Alfred, Lord Tennyson*

———

Evolutionary Humanism

Fifty years before Edward O. Wilson presented a sociobiological view of religion, Julian Huxley was exploring a similar path.

Julian Huxley

Religion in the light of science is seen not as a divine revelation, but as a function of human nature. It is a very peculiar and very complicated

function of human nature, sometimes noble, sometimes hateful, sometimes intensely valuable, sometimes a bar to individual or social progress.
But it is no more and no less a function of human nature than fighting or
falling in love, than law or literature.

*Because religious sentiment was so much a part of human nature, Huxley
foresaw grave difficulties for individuals whose outlooks had been shaped
by the scientific ethos.*

New ideas, usually classed as scientific, have permeated a large section of
the community and prevented them from belonging to any of the established churches, whose belief in miracles, in revelation, in the inspired
authority of the Bible, runs counter to the established truth, as the scientifically trained see it. The problem is to make a religion for these men
and women, whose numbers are bound to increase with the spread of
education, and who will otherwise be left without a religion, or with one
to which they cannot whole-heartedly give their assent. The conflict between religion and science in the last half-century resulted in the complete defeat of religion's claim to impose its view as authoritative on
man's mind, but it did not build up anything for those whom it emanci-
pated. That reconstruction is our problem today.

*Huxley recounts why, at the age of twenty-eight (and while chairing the
biology department at the new Rice Institute in Texas), he decided to
play a role in the birth of a religion that would square with science.*

Browsing the public library at Colorado Springs, under the shadow of
Pike's Peak, while waiting to go into hospital for an operation, I came
across some essays of Lord Morley in which there occurred the words,
"The next great task of science will be to create a religion for humanity."
I was impressed that a man of Morley's intellectual power and rationalis-
ing tendencies should have been so much interested in a religion for
humanity; I was fired by sharing his conviction that science would of
necessity play an essential part in framing any religion of the future worthy the name; and I was impressed too with his use of the impersonal
word 'science', as implying that any real progress in religion nowadays
would be the slow product of generations of thinkers and workers reacting on the common thought and practice of the times, much more than
the creation of a single personality, in this respect reversing the historical
process which had seen the traditional and communal religions of primitive peoples give place to the great historical religions—Buddhism, Christianity, and Islam—with their individual founders.

Morley's words made the more impression upon me, since already I
had conceived some half-hearted idea of attempting to restate the realities
of spiritual values which my experiences had forced upon me in terms of
an intellectual framework drawn from my scientific training; I was aiming at a harmony which, although only vaguely perceived, I yet felt must

exist and, if it existed, and could be found, would not only bring satisfaction to myself, but might save others from some of the conflicts and pains which I had been through. Morley's words confirmed me in my resolve to try to contribute to the task he envisaged.

How that religion will take form—what rituals or celebrations it might practise, whether it will equip itself with any sort of professional body or priesthood, what buildings it will erect, what symbols it will adopt—that is something which no one can prophesy. Certainly it is not a field on which the natural scientist should venture. What the scientist can do is to draw attention to the relevant facts revealed by scientific discovery, and to their implications and those of the scientific method. He can aid in the building up of a fuller and more accurate picture of reality in general and of human destiny in particular, secure in the knowledge that in so doing he is contributing to humanity's advance, and helping to make possible the emergence of a more universal and more adequate religion.

Huxley proffered "evolutionary humanism" as a candidate religion for the scientific age. The core belief of evolutionary humanism, which would draw forth the religious sentiments of its adherents, would be this:

Man is that part of reality in which and through which the cosmic process has become conscious and has begun to comprehend itself. His supreme task is to increase that conscious comprehension and to apply it as fully as possible to guide the course of events. In other words, his role is to discover his destiny as an agent of the evolutionary process.

Julian Huxley thus set himself the task of providing the intellectual roots of a new religion, from which he hoped might emerge rituals and other trappings of religious practice. Both cogent and literary, Huxley's appeal for a scientistic rendering of religion built upon an evolutionary consciousness is unsurpassed to this day. Excerpts here are drawn from his book Religion Without Revelation, *initially published in 1927. I have used a much later, revised edition (copyright 1957 by Julian Huxley), reprinted by permission of HarperCollins Publishers Inc. and the Huxley Estate (Peters, Fraser and Dunlop).*

Religion Without Revelation

Julian Huxley

I believe, in the first instance, that it is necessary to believe something. Complete scepticism does not work. On the other hand, I believe equally strongly that it is always undesirable and often harmful to believe without proper evidence. Religion of the highest and fullest character can coexist with a complete absence of belief in revelation in any straightforward sense of the word, and of belief in that kernel of revealed religion, a personal god.

I believe firmly that the scientific method, although slow and never claiming to lead to complete truth, is the only method which in the long run will give satisfactory foundations for beliefs. The scientific method is the method which, in the intellectual sphere, is the counterpart of that method recommended by the apostle in the moral sphere—test all things; hold fast to that which is good. It consists in demanding facts as the only basis for conclusions; and of consistently and continuously testing any conclusions which may have been reached by new facts and, wherever possible, by the crucial test of experiment. It consists also (and this is not sufficiently recognised by the generality of people) in full publication of the evidence on which conclusions are based, so that others may have the advantage of the facts, to assist them in new researches, or, as frequently occurs, to make it possible for them to put a quite different interpretation on the facts.

When there exists no evidence or next to no evidence, and when the conclusion to which we may come can have no influence on the facts, then it is our duty to suspend judgment and hold no belief, just as definitely as it is our duty, when practical issues hang on our decision, not to suspend judgment, but to take our courage in both hands and act on the best belief at which we can arrive. This duty of refraining from belief is often imposed upon men of science in their work, in order that they may in the long run arrive at greater certitude; it is also imposed upon them in other cases in order that they may not encourage false hopes of certitude. When applied to whole problems, this attitude of mind generally goes by the name (first coined by Thomas Huxley) of agnosticism. I hold it to be an important duty to know when to be agnostic. I believe that one should be agnostic when belief one way or the other is mere idle speculation, incapable of verification; when belief is held merely to gratify desires, however deep-seated, and not because it is forced on us by evidence; and when belief may be taken by others to be more firmly grounded than it really is, and so come to encourage false hopes or wrong attitudes of mind.

Our beliefs about Belief are among the most important that we may possess, and this all the more since we rarely stop to give their existence a thought. I hold, then, that all our life long we are oscillating between conviction and caution, faith and agnosticism, belief and suspension of belief. That neither faith nor agnosticism is in itself the better way, but that each has its right occasions. That beliefs which are well enough for individual occasions of practical necessity may be wholly unjustified and unjustifiable when made general or when taken to dispense from further enquiry.

I submit that the discoveries of physiology, general biology, and psychology not only make possible, but necessitate, a naturalistic hypothesis, in which there is no room for the supernatural, and the spiritual forces at work in the cosmos are seen as part of nature just as much as the material forces. What is more, these spiritual forces are one particular product of mental activity in the broad sense, and mental activities in general are seen to have increased in intensity and importance during the course of cosmic time. Our basic hypothesis is thus not merely naturalistic as opposed to supernaturalist, but monistic as opposed to dualistic, and evolutionary as opposed to static.

The time is ripe for the dethronement of gods from the dominant position in our interpretation of destiny, in favor of a naturalistic type of belief-system. The supernatural is being swept out of the universe in the flood of new knowledge of what is natural. It will soon be as impossible for an intelligent, educated man or woman to believe in a god as it is now to believe that the earth is flat, that flies can be spontaneously generated, that disease is a divine punishment, or that death is always due to witchcraft. Gods will doubtless survive, sometimes under the protection of vested interests, or in the shelter of lazy minds, or as puppets used by politicians, or as refuge for unhappy and ignorant souls: but the god type will have ceased to be dominant in man's ideological evolution.

A Feeling for the Sacred

It is frequently taken for granted that religion is essentially a belief in a god or gods. And yet, one of the great religions of the world, namely Buddhism, in its original and purest form does not profess belief in any supernatural being. What, then, is religion? It is a way of life. It is a way of life which follows necessarily from a man's holding certain things in reverence, from his feeling and believing them to be sacred. And those things which are held sacred by religion primarily concern human destiny and the forces with which it comes into contact.

The essence of religion springs from man's capacity for awe and reverence. The objects of religion, however much later rationalised by intellect or moralised by ethics, however fossilised by convention or degraded by superstition or fear, are in origin and essence those things, events, and ideas which arouse the feeling of sacredness. I believe, then, that religion arose as a feeling of the sacred. The capacity for experiencing this feeling seems to be a fundamental capacity of man, something given in and by the construction of the normal human mind.

Man's idea of the divine, and his expression of it, is on a par with his discovery and formulation of intellectual truth, his apprehension and expression of beauty, his perception and his practice of moral laws. There is no revelation concerned in it more than the revelation concerned in scientific discovery, no different kind of inspiration in the Bible from that in Shelley's poetry. That is to say that there is no literal revelation, no literal inspiration; and it is mere prevarication to shift, as is often done, from one sense of these words to the other, from the wholly literal, implying revelation or inspiration by supernatural beings, to the descriptive-metaphorical, implying only the flashing into consciousness of something new, independent of the will, and carrying with it a quality of essential rightness.

Here is a mass of a few kilograms, of substance that is indivisibly one (both its matter and spirit), by nature and by origin, with the rest of the universe, which can weigh the sun and measure light's speed, which can harness the tides and organise the electric forces of matter to its profit, which is not content with huts or shelters, but must build Chartres or the Parthenon; which can transform sexual desire into the love of a Dante for his Beatrice; which can not only be raised to ineffable heights at the sight of natural beauty or find 'thoughts too deep for tears' in a common flower, but can create new realms and even heavens of its own, through music, poetry, and art; which is never content with the actual, and lives not by bread alone; which is always not only surmounting what it thought were the limitations of its nature, but, in individual and social development alike, transcending its own nature and emerging in newness of achievement.

   We men are from one point of view mere trivial microbes, but from another the crown of creation: both views are true, and we must hold them together, interpenetrating, in our thought. From the point of view of the stellar universe, whose size and meaningless spaces baffle comprehension and belief, man may appear a mere nothing, and all his efforts destined to disappear like the web of a spider brushed down from the corner of a little room in the basement of a palace; but meanwhile he is engaged upon a task which is the most valuable of any known, the most valuable which by him can be imagined, the task of imposing mind and spirit upon matter and outer force. This he does by confronting the chaos of outer happenings with his intellect, and generating ordered knowledge; with his aesthetic sense, and generating beauty; with his purpose, and generating control of nature; with his ethical sense and his sense of humour, and generating character; with his reverence, and generating religion.

The Need for Change

Religious ideas and practices may be, as in many primitive peoples, closely adapted to the general life of society: when civilisation is rapidly changing, however, they are often either ahead of or behind the general thought of the time. The ethics and spiritual insight of Jesus and of Buddha, for instance, were far ahead of their times, as was the theological insight of Abelard, or the moral zeal of some of the Hebrew prophets, or the love of learning by the better of the monks in the Dark and early Middle Ages. More often, however, the unfortunate tendency of the sacred to become the untouchable, and therefore for religion to become an unduly conservative force, has often led to religious thought and practice

being below the general level of its times. The anti-evolution agitation in this country a hundred years ago, and in the United States in the present century, is an instance in the intellectual sphere; the refusal of the Roman Catholic and other churches to discuss such subjects as divorce and birth control in any reasonable spirit is an example in the moral sphere; the excessive formalism of orthodox Jewish religion in the time of Jesus is an example in the field of ritual; the intolerance by many missionary societies of native custom and belief, as compared with the views of anthropological science and of enlightened administrators, will serve in the field of social ethics.

In all spiritual activities we should expect steady change and improvement as man accumulates experience and perfects his mental tools. As matters stand today, we have the cleavage between orthodox religion which has hitched its wagon—the sense of the sacred—not to a star, but to a traditional theology, and a large body of educated people who, rejecting the theology, are forced to stand outside religion too.

In art it is a triumph if a Beethoven or a Titian finds new ways of building beauty; in science it is acclaimed a triumph if an old universally accepted theory is dethroned to make way for one more comprehensive, as when Newton's mechanics gave place to Einstein's, or the assumed indivisibility of the atom was exploded in favour of the compound atom, organised out of subatomic particles; but in the religious sphere, owing largely to this pernicious view that religion is the result of supernatural revelation and embodies god-given and therefore complete or absolute truth, the reverse is the case, and change, even progressive change, is by the great body of religiously-minded people looked upon as a defeat. Whereas once it is realised that religious truth is the product of human mind and therefore as incomplete as scientific truth, as partial as artistic expression, the proof or even the suggestion of inadequacy would be welcomed as a means to arriving at a fuller truth and an expression more complete.

Earlier religions and belief-systems were largely adaptations to cope with man's ignorance and fears, with the result that they came to concern themselves primarily with stability of attitude. But the need today is for a belief-system adapted to cope with his knowledge and his creative possibilities; and this implies the capacity to meet, inspire and guide change. In other words, the primary function of earlier systems was of necessity to maintain social and spiritual morale in face of the unknown: and this they accomplished with a considerable measure of success. But the primary function of any system today must be to utilise all available knowledge in giving guidance and encouragement for the continuing adventure of human development.

I am here treating religious systems as social organs whose function it is to adjust man to his destiny. No previous systems could perform this function with full adequacy, for the simple reason that no previous age had sufficient knowledge to construct an adequate picture of the drama of destiny or of its protagonist, man. The present epoch is the first in which such a picture could begin to take shape. Immense tracts of ignorance are still to be explored, and await annexation to the growing empire of knowledge; but we can already affirm that the cosmos is unitary, that it is a process of transformation in time, and that values and other products of mental and spiritual activity play an important operative role in that sector of the process with which we are involved.

More specifically, the present is the first period in the long history of the earth in which the evolutionary process, through the instrumentality of man, has taken the first step towards self-consciousness. In becoming aware of his own destiny, man has become aware of that of the entire evolutionary process on this planet: the two are interlocked. This is at once an inspiring and a sobering conception.

The present age also differs from all earlier ages in the increased importance of science, and its universal extension. There should no longer be any talk of conflict between science and religion. Between scientific knowledge and certain religious systems, yes: but between science as increasing knowledge of nature and religion as a social organ concerned with destiny, no. On the contrary, religion must now ally itself wholeheartedly with science.

Science in the broad sense is indispensable as the chief instrument for increasing our store of organised knowledge and understanding. Through evolutionary biology, it has already indicated the nature of human destiny. Scientific study is needed to give religion a fuller understanding of destiny, and to help in devising better methods for its detailed realisation. The contribution which science can make is two-fold. It can contribute an enormous body of hard-won, tested, organised knowledge; and also a spirit of disinterested devotion to truth, and a willingness to apply this spirit to any problem, irrespective of prejudices or possible consequences.

It was impossible, before the later nineteenth century, to have any properly-grounded idea of the unity of the natural world. Such ideas had been indulged in, but were rightly treated as hazardous speculations: now, they are forced upon our minds by the irresistible body of tested knowledge. In the same way, there is the sense of solidity given by the mere detailed knowledge and comprehension of the facts of nature—how the wind blows and the clouds arise; how valleys and mountains obtain their forms and gradually change; how the sun shines and the earth and moon circle round it; why earthquakes and volcanic eruptions take place; the how of chemical combination and the knowledge of the composition and properties of familiar substances; the way in which we and animals breathe, digest, move, feel, reproduce; how diseases are caused by bacteria, parasites, definite poisons or lack of definite food-substances—in all these and a hundred other ways civilisation has an assured sense of acquaintance, a foundation of knowledge built in the world of external nature, which was impossible to any previous age.

How can the twentieth century, grounded in this outlook, which is not only actually but inevitably new, be content with the same religious outlook which satisfied it when the natural world was uncomprehended, appeared chaotic as much as orderly, and the ideas of control and conscious change had not yet been born?

Books Instead of Sermons

If the spiritual life-blood of the great masters of thought is available to everyone, why go to church and listen to familiar prayers and to a prosy sermon, when you could stay at home and receive new knowledge and deeper thoughts from a book? Goethe, Emerson, Wordsworth, Blake, Carlyle, Dante, Sir Thomas Browne, Shelley, and the rest of the assembly of immortal spirits—they jostle each other on your shelves, each waiting only to be picked up to introduce you to his own unique and intense experience of reality.

The *Origin of Species* is today a good deal more profitable as theology than the first chapter of Genesis, and William James' *Principles of Psychology* will be a better commentary on the Decalogue than any hortatory sermon. The poetry of Herbert or Donne or Vaughan, of Francis Thompson or Walt Whitman, will introduce you to new ways of mystic feeling; Trevelyan's *History of England* is likely to be a more salutary history lesson, because nearer home, than the historical books of the Old Testament; Whitehead's *Science and the Modern World* is more likely to help the perplexed mind of a twentieth-century Englishman than the apocalyptic visions of Revelations or the neo-Platonic philosophy of the Fourth Gospel; to sacrifice a score of Sundays to making acquaintance with the ideas of other great religions like Buddhism might be very much preferable, even from the purely religious point of view, to continuance in the familiar round and the familiar narrowness of one's own church.

And the same is true in other spheres. You may get much more spiritual exaltation and joy out of a Beethoven concert, or even out of Beethoven on the gramophone, than by listening to your local organist play Mendelssohn or the 'March of the Priests.' You may exercise your highest faculties by travel, now that travel is easy and cheap, or you may stay at home and discipline your mind with reproductions of what the great artists and sculptors and architects have imagined and expressed. If you are primarily in love with morality and character, you may obtain a fuller insight into them by reading the great novelists and playwrights, and the biographies of great men, than by confining yourself to the Bible and the saints.

What is more, there is no reason whatever why in all such activity you should not in your degree be participating in the religious life. All philosophy and science, all great art, all history, all lives of men—one and all may inspire to reverence or exaltation, or be made the subject of reflection which, being concerned with great problems of destiny in a grave and reverent way, is more truly religious than any pietism.

It might be said that there is no room left for organised worship. I do not think this is so. There will always remain the religious satisfaction of plunging the mind in a common social act, and always a satisfaction in a familiar ritual hallowed by time and association. There is also to many people a satisfaction in symbolism; and to others in finding, in the combined privacy and publicity of the church service, a simultaneous release from the world and from the individual self.

Religion Without Divinity

There does exist an outer ground and object of religion as much as an outer object of science. The fact, however, that this outer object is by most religions considered to be an external divine being is, philosophically speaking, an accident. If, however, this superposed belief and its corollaries be removed, what remains of the reality?

The answer is 'a great deal.' That reality includes permanent facts of human existence—birth, marriage, reproduction, and death; suffering, mutual aid, comradeship, physical and moral growth. It includes also other facts which we may call the facts of the spiritual life, such as the conviction of sin, the desire for righteousness, the sense of absolution, the peace of communion; and those other facts, the existence and potency of human ideals, which, like truth and virtue and beauty, always transcend the concrete and always reveal further goals to the actual. It also includes facts and forces of nature outside and apart from man—the existence of matter and of myriads of other living beings, the position of man on a little planet of one of a million suns, the facts and laws of motion, matter and energy and all their manifestations, the history of life.

I say that it includes these; it would be more correct to say that it includes certain aspects of all these and many other facts. It includes them in their aspects of relatedness to human destiny; and it includes them as held together, against the cosmic background, by a spirit of awe or reverence. If you wish more precision, it includes them in their sacred aspect, or at least in association with an outlook which is reverent or finds holiness in reality. Finally, it includes them not merely disjointedly, as so many separate items: it includes them in a more or less unified whole.

For my own part, the sense of spiritual relief which comes from rejecting the idea of God as a supernatural being is enormous. I see no other way of bridging the gap between the religious and the scientific approach to reality. But if this rejection is once accomplished, the abyss has disappeared in the twinkling of an eye, and yet all the vital realities of both sides are preserved. The mental life of humanity is no longer a civil war but a corporate civilisation. Within it there will be conflicts, frictions, adjustments; but these are inevitable, and probably necessary for full vitality, and if they take place within a whole which is organised for unity and production instead of duality and strife, there will be advance.

A developed religion must satisfy the following requirements. It will not merely be confined to man's more or less immediate reaction to the mysterious or sacred; it will not be content with a system (often incomplete or self-contradictory) of mythology or of primitive rationalisation as its theology; nor only with traditional ritual or formalism as its code of action. On the contrary, it will always extend its conception of what is sacred and a proper object of religious feeling to include man's destiny and his relation with the rest of the world; it will apply the pure force of intellect to its ideas, and attempt a theology or intellectual basis which shall be both logical and comprehensive, accurate and coherent; it will also inevitably perceive that ethics and morality are keystones of human destiny, and link up its sacred beliefs with a pure ethic and a reasoned morality. It will, in a word, not be content to leave its religious life chaotic and unordered, with loose ends unconnected with the rest of reality, but will come more and more consciously to aim at an organised and unified scheme of religion, which further shall be connected with all other parts of the mental life; and it will attempt to achieve this by putting forward a scheme of belief and a scale of values around and over which man's aspirations to sacredness in emotion, thought, and action may most securely grow.

Thus a developed religion should definitely be a relation of the personality as a whole to the rest of the universe, one into which reverence enters, and one in which the search for the ultimate satisfactions of discovering and knowing truth, experiencing and expressing beauty, and ensuing the good in righteous action, all have the freest possible play.

The Need for Tolerance

If our religion is a true religion, a religion of fuller life, it must both tolerate and reverence variety. The efficient biologist or engineer who would deny all value to religious meditation and the religious life; the missionary who begins by suppressing all native activities of which he in the least disapproves; the scholastic theologians who denied independent value to natural science or humanist philosophy; the efficient administrator who would lock up as a vagrant everyone who is not constantly at work—all are limited in their outlook, and because limited therefore wrong. Even Plato, at the full flight of his imagination, desiring to banish poetry and art from his ideal republic, was subordinating reality to logic and had failed to gain a full vision of truth and virtue.

But our tolerance must not be merely passive, a tired intellectual gesture; it must be active, springing from the belief and knowledge that truth is too large to be revealed in but one form, or one creed, or one way of life. We must accept the hard saying that out of diversity alone comes advance, and that any one human mind is too small to grasp more than a little truth, to live more than a little reality.

I feel that any religion of the future must have as its basis the consciousness of sanctity in existence—in common things, in events of human life, in the gradually-comprehended interlocking whole revealed to the human desire for knowledge, in the benedictions of beauty and love, in the catharsis, the sacred purging, of the moral drama in which character is pitted against fate and even deepest tragedy may uplift the mind. It must admit that this same high sense of sacredness and transcendent value may be vouchsafed in many ways and in many objects. Some may find it in poetry. Shelley was an avowed atheist and a hater of Christianity, but he was obviously of religious temperament. This could hardly be claimed of Keats, but to him beauty was certainly a sacrament. It may come through art or music; it may be vouchsafed through love. It may be found in the pursuit of pure truth—think of Lucretius, Galileo, Pasteur, Thomas Huxley. It may be found in the practice of a life devoted to the service of humanity's suffering, as with Father Damien, or Mrs Elizabeth Fry, or Dr Schweitzer. Still others, like Richard Jefferies or Wordsworth or Thoreau, may find it in the solitudes of nature; or, again, like born patriots, in a sanctification of their country.

Man's scale of desires and values, his spiritual capacities, dictate the direction of his religion, the goal towards which it aspires; the facts of nature and life dictate the limits within which it may move, the trellis on whose framework those desires and emotions must grow if they are to receive the beams of truth's sun, if they aspire above creeping on the ground. It is our duty to know those outer facts truly and completely, to be willing to face all truth and not try to reject what does not tally with our desires: and it is our duty to realise our own capacities, to know what desires are to be put in command, what desires are to be harnessed to subordinate toil, to place our whole tumultuous life of feeling and will under the joint guidance of reverence and reason.

In so far as we do this, we prevent the man of devout religious feeling from being subordinated to a system which may organise the spirit of religion in opposition to discovery or necessary change, or may discharge its power in cruelty and persecution; and we help religion to help the progress of civilisation. But in so far as we neglect this, we are making man a house divided against itself, and allowing the strong tides of religious feeling to run to waste or to break in and devastate the fruit of man's labour. And the choice is in our own hands.

———

I am the credulous man of qualities, ages, races;
I advance from the people in their own spirit;
Here is what sings unrestricted faith.

I, too, following many, and follow'd by many, inaugurate a
    Religion—I descend into the arena.

Each is not for its sake,
I say the whole earth, and all the stars in the sky,
    are for Religion's sake.

I say no man has ever yet been half devout enough;
None has ever yet adored or worship'd half enough;
None has begun to think how divine he himself is,
    and how certain the future is.

O strain, musical, flowing through the ages—now reaching hither!
I take to your reckless and composite chords—I add to them,
    and cheerfully pass them forward.

—*Walt Whitman*

———

Darwinism with a Difference

E. O. Wilson and Julian Huxley call for extension of evolutionary biology into the realm of religion. Wilson's plea for evolution as religion is rational and pragmatic. Evolution is, quite simply, "the best myth we will ever have." Such extension is, however, a task for others; Wilson gives no hint that he personally has fallen to the power of deep-seated, genetic drives for religious sentiments. Julian Huxley, on the other hand, has carried out an extension. His "evolutionary humanism" is, moreover, heartfelt; Huxley is unquestionably a believer. But Huxley is not a believer in a theistic, deistic, or even immanent god. Divinity for Huxley is found in humankind and in the evolutionary process that stretches both behind and ahead of us.

    Does a grasp of evolutionary biology preclude more traditional religious sentiments? The life and writings of Alister Hardy demonstrate that not all extensions of evolutionary biology into the realm of religion need conflict with belief in a greater power, even a personal god. Hardy was both a distinguished zoologist and (later in life) an investigator of religious experience. Chapter 5 presented passages from one of his scientific books, The Living Stream. *He also wrote* The Divine Flame, The Biology of God, *and* The Spiritual Nature of Man. *Brief selections that follow are drawn from his last book,* Darwin and the Spirit of Man, *which was published in* 1984, *one year before his death.*

Alister Hardy

What an extraordinary position it is that the intellectual world is now finding itself to be in. A large section of the public, including some of the best brains, has been convinced by the work of the molecular biologists that the neo-Darwinian doctrine of today is pointing only to a materialistic interpretation of life. I am myself an ardent neo-Darwinian, but I must also admit that all my life I have been one of those who has had a sense of spiritual awareness. I should, however, make clear that I do not believe in God as a kind of old gentleman out there; that image, as Bishop John Robinson said in his *Honest to God* (1963), must go. Nevertheless I am sure that our relationship with what we call God must be a personal one; it must be an intimate I-Thou relationship, in fact a deep love relationship.

I have already indicated that an animal's power of choice—conscious choice I believe—can act by behavioural selection within the Darwinian system. The various organs of the body, like the hands and feet, I have said, whilst fully described in terms of physics and chemistry, have been selected to a larger and larger degree among the higher animals by the conscious behaviour of the animals themselves. How am I to link this conscious behaviour, which works within the Darwinian system by means of its particular form of selection, to the subsequent development of religion and the spiritual nature of man? In company with some of the more philosophical biologists such as Sir Charles Sherrington, Sir John Eccles, Sir Cyril Hinshelwood and philosophers such as A. N. Whitehead and Sir Karl Popper, I must admit to being a dualist, i.e. one who believes that the universe has at least a double nature, being not entirely material, but having an equally important mental and spiritual element.

If we are attempting to link man's spiritual nature with evolution, are we to suppose that animals may have some form of experience coming from somewhere other than through the organs of sense? It would appear to be most unlikely, and anyway we could never know. However, when the change came in evolution, we can see that primitive man, with his new powers of communication by language and the handing on of tradition as a result of shared experience, could develop feelings of a new awareness. They could build up a general tradition of there being some element with which they could make contact; and if they approached it with a certain reverence, with a devotional feeling, with in fact a form of love akin to that between child and parent, they would in return feel a lifting up of the self, a new sense of confidence, a power to overcome difficulties, a force to make them stronger, a something that gave more courage than they had ever had before. They might call it *mana, waken, nhialic, knoth,* and other names; and later it could be called God. We can in fact see its link with evolution; it would have, in the words of

Marett, "survival value" in that those primitive tribes which developed more courage, felt themselves receiving this support, would more likely be successful in the struggle for life.

I am often asked "Can anything be said about a possible purpose in the universe?" I think most agnostics and humanists, if asked, would say "No." Now I admit that I don't think anything useful can be said, because it can be nothing more than the wildest possible speculation; nevertheless perhaps it might just be worth saying that, on purely logical grounds, it is not impossible to imagine a reasonable goal for the cosmic evolutionary process. Certainly any such guess made in the twentieth century is most unlikely to be the correct one; however, the very fact that one can conceive an even remotely possible solution may save one from being in the pessimistic position of imagining that there can be no possible purpose in the process at all. Such a defeatist loss of all sense of meaning in the world is one of the tragic outcomes of the materialism of today.

My tentative and no doubt entirely improbable answer cannot be other than a quite fantastic flight of fancy into the realm of science fiction. Perhaps I am making a great mistake and am simply making myself ridiculous. I take the risk of this because without outlining the idea, which I agree is no doubt almost absurd, one cannot vindicate one's belief that logically it is not impossible to conceive of a cosmic plan in the evolutionary process. So with this warning of what nonsense to expect, I will apologetically proceed.

We see the continually increasing rate of man's scientific and technological skill and achievements. The progress in the present century is staggering. All this great development has come about in an extraordinary short space of time and indeed the whole of man's civilization is but a few thousand years compared with the two thousand million years span of organic evolution. Provided we have no cosmic or man-made disaster, we should, on this earth alone, still have more than a thousand million years of evolution in front of us.

Given sufficient time and a sufficient increase in information—stored mechanically for reference far beyond the memories of individual men—it seems possible that there must logically come a point when man has indeed asked every possible question that can be asked and has in time gained and recorded every possible answer. He will know all the secrets of the universe. He may in addition have developed a new collective consciousness and a greatly increased spirituality.

I am, perhaps, in my imaginary answer getting near the Omega point of Teilhard de Chardin, but by a different road. Perhaps indeed we really are the children of God and that evolution must, with its spiritual element operating within the material matrix, eventually lead to a collective omniscient consciousness knowing just how and where in the universe

life may and will be started again! We are perhaps part of a great system for generating love, joy and beauty in the universe: the highlights of existence that can only be perceived and appreciated when seen against the darker background of their opposites.

Yes, to imagine a purpose is not impossible; but to suppose that what we imagine is actually the real purpose would be the height of impertinence. Let us leave it at that and return to earth; we may return full of confidence that, from what the mystics, poets and artists tell us, the real meaning of the cosmic process is something far more wonderful than anything we can possibly imagine in our present state of being.

———

Religion is the vision of something which stands
Beyond, behind, and within the passing
Flux of immediate things;
Something which is real, and yet waiting
To be realised; something that gives meaning
To all that passes, and yet eludes apprehension;
Something whose possession is the final good,
And yet is beyond all reach;
Something which is the ultimate ideal,
And the hopeless quest.

—*Alfred North Whitehead*

———

Opposing Evolutionary Extensions

All these forays into speculative philosophy and religion by evolutionary biologists have not gone unchallenged. John C. Greene, one of this century's most respected historians of science, is a powerful critic and has devoted a great deal of attention to the worldviews emanating from evolutionary biology. In a 1981 book, Science, Ideology, and World View, *Greene launches a spirited assault on extensions of evolutionary biology into religious realms. He upbraids a number of transgressors, including Julian Huxley. Even George Gaylord Simpson, who in most of his writings gives formal religion wide berth, is chastised by Greene for attempting to find any meaning at all in the fossil record. Selections from* Science, Ideology, and World View: Essays in the History of Evolutionary Ideas *by John C. Greene (copyright 1981 by the Regents of the University of California) are reprinted by permission of the author and University of California Press.*

John C. Greene

What is the relevance of evolutionary biology to the question of human duty and destiny? And is the attempt of modern evolutionary biologists to provide answers to this question from their scientific findings not as futile and delusory as were the efforts of William Buckland, Adam Sedgwick, and Benjamin Silliman to demonstrate the existence and attributes of God from the record of the rocks?

In both cases, the conclusions antedate the investigation and dictate its outcome. Simpson and Huxley did not learn to value the individual, to detest totalitarianism, and to believe in the brotherhood of man from studying biology and paleontology any more than Buckland and his contemporaries learned to believe in an omnipotent, omniscient, and benevolent Creator from their scientific researches. One group of scientists approached nature as agnostics or atheists, the other as Christians; each found what it expected to find. Each had a moral commitment, the first to vindicate the values they held dear and discredit beliefs they regarded as inimical to the further progress of science, the second to confound the skeptics and corroborate religious doctrines that made human duty and destiny intelligible to them. In both cases science was only a tool, a weapon, in defense of positions that were essentially religious and philosophical. Faith in science can be no less a religion than faith in God.

To the historian of ideas it seems evident that champions of evolutionary biology as the clue to human duty and destiny are caught on the scientistic horn of the positivist dilemma. Whoever regards science as man's sole means of acquiring reliable knowledge of reality must eventually confront that dilemma. If science and the scientific method are defined narrowly so as to exclude value judgments, it then becomes impossible to say why anything, science included, is important or valuable, why the passion for truth is to be inculcated and respected, or why human beings have any more inherent dignity than starfish or stones. But if, on the contrary, science is declared competent to discover human duty and destiny, as those who choose the other horn of the dilemma assert, one is soon confronted with the conflicting claims of Huxleyan science, Darlingtonian science, Freudian science, Marxian science, Comtean science, and a host of other scientisms. In the ensuing struggle the central idea of science as an enterprise in which all qualified observers can agree as to what the evidence proves vanishes from sight. Thus, whichever horn of the dilemma the positivist takes, science is the loser.

In any case, it seems clear that with respect to the fundamental problems of philosophy the post-Darwinian writers are, as T. H. Huxley said long ago, "exactly where the pre-Darwinian generations were." What is nature? What is mind? How are they related? What is man? If man is part and parcel of nature, must not nature in some sense be like man? But if man can comprehend nature and set himself against it or direct its course, must he not transcend nature? And how can nature be comprehensible to man's mind if that mind is totally unlike it?

These are old questions, as old as philosophy itself, and from what we have seen of Darwinism as a world view it seems unlikely that evolutionary biology in and of itself will ever provide intelligible answers to them. Shall we not, then, resume the age-old quest for knowledge of reality in a humbler spirit, acknowledging our debt to science for all it can tell us about ourselves and nature but realizing that the ultimate intelligibility of things, if there is one, is not scientific in the sense that we understand that word today. If human life were to come to an end fifty or one hundred years from now through human pollution of the natural environment, the event would presumably have a scientific explanation. But would the explanation satisfy anyone—even Sir Julian Huxley?

———

They tell me there are leucocytes
In my blood and sodium and carbon in my flesh.
I thank them for the information and tell them
There are black beetles in my kitchen,
Washing soda in the laundry and coal in my cellar.
I do not deny their existence
But I keep them in their proper place.

—*George Bernard Shaw*

Charles Sanders
Pierce

The *Origin of Species* of Darwin merely extends politico-economical views of progress to the entire realm of animal and vegetable life. As Darwin puts it on his title-page, it is the struggle for existence; and he should have added for his motto: Every individual for himself, and the Devil take the hindmost! Jesus, in his sermon on the Mount, expressed a different opinion. Here, then, is the issue. The gospel of Christ says that progress comes from every individual merging his individuality in sympathy with his neighbors. On the other side, the conviction of the nineteenth century is that progress takes place by virtue of every individual's striving for himself with all his might and trampling his neighbor under foot whenever he gets a chance to do so. This may accurately be called the Gospel of Greed.

Charles Sanders Pierce, one of the founders of pragmatism (that quintessentially American philosophy), published the above rant against Darwinism in 1893. Notably, Pierce faults the Darwinian worldview for an utterly pragmatic reason: he objects to the support it lends to the "Gospel of Greed." Nineteenth-century presentations of the theory of evolution by natural selection did tend to employ Darwin's own brutal metaphors; even textbooks contained blatantly value-laden passages. But, as evident in earlier chapters of this book, evolutionary theorists of the late twentieth century are doggedly pluralistic in their interpretations and their extensions. Moreover, high school texts now put the idea of evolution into rather tepid (indeed, far-from-memorable) terms.

Why, then, the fuss? Why is there a push for teaching "creation science," or its more current incarnations as "abrupt appearance theory" and "intelligent design theory," with, and as an alternative to, evolutionary biology in public schools in the United States?

The Crusade for Creation Science

Henry M. Morris

The Evolution Model, by its very nature, is an atheistic model (even though not all evolutionists are atheists) since it purports to explain everything without God. The Creation Model, by *its* nature, is a theistic model (even though not all creationists believe in a personal God) since it requires a creator able to create the whole cosmos. The Creation Model is at least as scientific as the Evolution Model, and evolutionism is at least as religious as creationism. Theism and atheism are mutually exclusive philosophies and are therefore in the same category. It is not more nonreligious for a view to be atheistic than to be theistic.

This passage appears in the introduction to the book What Is Creation Science? *by Henry M. Morris and Gary E. Parker (first edition published in 1982). Morris is, perhaps, the most widely known proponent of the scientific merits of a creationist view of origins; he is also a leader of the social movement that is pushing for "balanced treatment" of creationist with evolutionist arguments in public school science. Morris came to his ideas not through an unquestioning allegiance to a creed acquired right out of the cradle; his convictions owe to a personal quest he took on as a young adult. The biographical excerpt on Morris that follows is drawn from "Creationism in Twentieth-Century America" by Ronald L. Numbers; it appeared in a 1982 issue of the journal* Science.

Ronald L. Numbers

Reared a nominal Southern Baptist, and a believer in creation, Morris as a youth had drifted unthinkingly into evolutionism and religious indifference. A thorough study of the Bible after his graduation from college convinced him of its absolute truth and prompted him to re-evaluate his belief in evolution. After an intense period of soul-searching he concluded that creation had taken place in six literal days because the Bible clearly said so and "God doesn't lie." Corroborating evidence soon came from the book of nature. While sitting in his office at Rice Institute, where he was teaching civil engineering, he would study the butterflies and wasps that flew in through the window; being familiar with structural design, he calculated the improbability of the development of such complex creatures by chance. Nature as well as the Bible seemed to argue for creation.

Henry Morris later earned a Ph.D. in hydraulic engineering and went on to chair the civil engineering department of Virginia Polytechnic Institute. He is now president of the Institute for Creation Research in Santee, California. The institute, which offers master's degrees in science, had been threatened with nonrenewal of its teaching license. Convinced that the State of California was acting in a prejudicial manner, the institute took the issue to court and in 1992 was awarded $225,000 in an out-of-court settlement. The settlement stipulated that "a private postsecondary educational institute may teach the creation model as being correct provided that the institution also teaches evolution." Also stipulated was that a government agency has no jurisdiction over the curriculum, course content, or philosophy of any private religious school.

That win for the creationist camp was offset the same year by new guidelines that went into effect for public schools in California. The State board of education decided that evolution should be given a central place in the science curriculum and that the social science curriculum is, rather,

the proper place to explore creationism as an alternative to evolutionism, should a school wish to do so. "Creation science is not a science and has no place in the science curriculum," concluded Dan Chernow, chairperson of the curriculum commission that advised the board.

Many of the leaders of the creation science movement, like Henry Morris, are Christians of fundamentalist persuasion. Their belief in the inerrancy of the Bible and the rightness of a literal interpretation of Genesis underlies their zeal. In a tract written for fellow fundamentalists, Morris states:

Henry M. Morris

It is obvious that one can logically reject the historicity of Genesis I–II only if he likewise rejects the rest of the Bible as well, and even the infallibility of Christ himself. Many modern-day religious liberals and even some supposedly conservative Christians have done exactly that. Most Christians, however, are unwilling to go this far. Some try to avoid the issue altogether, but this tactic almost inevitably is a prelude to compromise. The only Bible-honored conclusion is, of course, that Genesis I–II is the actual historical truth, regardless of any scientific or chronologic problems thereby entailed.

However one may judge the merit of Morris's personal motivation, the quality of his argument in behalf of creation science is a separate matter. This book will not, however, enter into the substantive debates. It will neither set forth nor critically examine the scientific arguments raised by creation advocates. The topic here is "evolution extended"—not "evolution questioned." Nevertheless, Henry Morris's argument that evolution is just as much a religion as is creationism is right on point with the theme of this chapter. The following excerpts are drawn from Morris's "Evolution as Religion" in the book What Is Creation Science? *by Henry M. Morris and Gary E. Parker. Copyright 1982/87 by Master Books (El Cajon, California), these passages are reprinted courtesy of Henry M. Morris.*

Evolution as Religion

It is an amazing thing that the modern establishments in science, education, and the news media continually portray creationism as religious and evolutionism as scientific. While the purpose of this book is to discuss only the scientific aspects of the two models, it is important also that readers at least be aware that evolutionism is much more 'religious' in essence than creationism. Evolution serves as the basic philosophy for many more religions of the world, past and present, than does special creation.

The following is a partial listing of those religions that are structured around an evolutionary philosophy: Buddhism, Hinduism, Confucianism, Taoism, Shintoism, Sikhism, Jainism, Animism, Spiritism, Occultism, Satanism, Theosophy, Bahaism, Mysticism, Liberal Judaism, Liberal Islam, Liberal Christianity, Unitarianism, Religious Science, Unity, Humanism. I am not claiming that all of these are based on modern Darwinism, for most of them antedate Charles Darwin. Nevertheless, they are all anti-creationist evolutionary religions, and have generally adapted easily to modern "evolution science."

The religions listed above are all extant religions, but the same discussion could apply to all the ancient pagan religions as well, all of which were essentially various forms of pantheism, and none of which were based on creation. Many of them (Epicurianism, Atomism, Stoicism, Gnosticism, pre-Confucian Chinese religions and many others) had cosmogonies quite similar to modern 'scientific' evolutionary cosmogonies. Most of them incorporated astrology, spiritism and idolatry into their systems as well.

Not only are the religions of atheism and humanism firmly grounded in evolutionary philosophy, but so also are a host of social, economic, and psychological systems which have had profound effect on human moral behavior and thus also are fundamentally religious. This includes such politico-economic systems as Marxism, Fascism and Nazism, and such psychological systems as Freudianism, behaviorism, and existentialism. It would include racism, imperialism, and laissez-faire capitalism on the one hand, and socialism, communism, and anarchism on the other.

The basic criterion of evolutionism is the rejection of a personal transcendent Creator who supernaturally called the space-time universe into existence out of nothing but His own omnipotence. All of the above religions regard the universe itself as eternal, constituting the only ultimate reality. Processes innate to the eternal space-time cosmos have developed the universe and its inhabitants into their present forms. These natural processes may, in many cases, be personified as gods and goddesses, but they are really just the natural processes innate to the universe itself. In some cases, the cosmos itself may be regarded as living and intelligent, giving rise not only to animals and people but also to 'spirits' who inhabit it. All of these concepts are evolutionary concepts, since none of the components or inhabitants of the universe are accepted as the products of fiat creation by an eternal Creator. The very existence of such a Creator is either denied or incorporated into the cosmos itself.

Thus, evolution is surely a religion, in every sense of the word. It is a world view, a philosophy of life and meaning, an attempt to explain the origin and development of everything, from elements to galaxies to people, without the necessity of an omnipotent, personal, transcendent Creator. It is the basic philosophy of almost all religions (except the few monotheistic religions), both ancient and modern. It is absurd for evolutionists to insist, as they often do, that evolution is science and creation is religious.

There are essentially only three modern creationist religions, in contrast to the dozens of evolutionary religions and religious philosophies. These are the monotheistic faiths—orthodox Judaism, orthodox Islam, and orthodox Christianity. These are all founded upon belief in one self-existent eternal Creator, who called the universe itself into existence in the beginning, as well as all its basic laws and systems.

Belief in this primeval special, completed, supernatural creation is consistent with all genuine facts of science, which is sufficient warrant for identifying this belief as "scientific creationism" or "creation science." This is further strengthened by the historical fact that most of the great scientists of the past who founded and developed the key disciplines of science were creationists. Note the following sample:

Physics: Newton, Faraday, Maxwell, Kelvin

Chemistry: Boyle, Dalton, Pascal, Ramsay

Biology: Ray, Linnaeus, Mendel, Pasteur

Geology: Steno, Woodward, Brewster, Agassiz

Astronomy: Kepler, Galileo, Herschel, Maunder

These men, as well as scores of others who could be mentioned, were all creationists, not evolutionists, and their names are practically synonymous with the rise of modern science. To them, the scientific enterprise was a high calling, one dedicated to "thinking God's thoughts after Him," as it were, certainly not something dedicated to destroying creationism.

New Religious Outgrowths of Evolutionism

A strange religion has been coming into prominence in recent years. Sometimes miscalled the 'New Age Movement', this phenomenon is in reality a complex of modern science and ancient paganism, featuring systems theory, computer science, and mathematical physics along with astrology, occultism, religious mysticism, and nature worship. Ostensibly offered as a reaction against sterile materialism of Western thought, this influential system appeals both to man's religious nature and his intellectual pride.

Although New-Agers have a form of religion, their 'god' is still Evolution, not the true God of creation. Many of them regard the controversial priest, Teilhard de Chardin, as their spiritual father. The ethnic religions of the East (Hinduism, Taoism, Buddhism, Confucianism, etc.), which in large measure continue the polytheistic pantheism of the ancient pagan religions, have long espoused evolutionary views of the universe and its living things, and so merge naturally and easily into the evolutionary framework of the New Age philosophy. It is surprising, however, to find that Julian Huxley and Theodosius Dobzhansky, the two most prominent of the western scientific neo-Darwinians, were really early proponents of this modern evolutionary religion.

Sir Julian Huxley, arguably the leading architect of the neo-Darwinian system, had written an influential book called *Religion without Revelation,* and had become, with John Dewey, a chief founder of the American Humanist Association. As first director-general of UNESCO, he formulated the principles of what he hoped would soon become the official religion of the world: "Thus the general philosophy of UNESCO should, it seems, be a scientific and world humanism, global in extent and evolutionary in background."

The neo-Darwinian religionists (Huxley, Dobzhansky, Dewey, etc.) thought that evolutionary gradualism would become the basis for the coming world humanistic religion. Evolutionists of the new generation, on the other hand, have increasingly turned to punctuationism—or revolutionary evolutionism—as the favored rationale, largely because of the scientific fallacies in gradualism increasingly exposed by creationists. This development has facilitated the amalgamation of Western scientism with Eastern mysticism:

The new systems biology shows that fluctuations are crucial in the dynamics of self-organization. They are the basis of order in the living world: ordered structures arise from rhythmic patterns. . . . The idea of fluctuations as the basis of order . . . is one of the major themes in all Taoist texts. The mutual interdependence of all aspects of reality and the nonlinear nature of its interconnections are emphasized throughout Eastern mysticism.

The author quoted, Dr. Fritjof Capra, at the University of California (Berkeley), is one of the New Age Movement's main scientific theoreticians, particularly in the application of modern computerized networking and systems analysis to the study of past and future evolution, also appropriating the unscientific idea of 'order through chaos', an ancient pagan notion reintroduced to modern thought by Ilya Prigogine.

"Gaia", the religion of the Earth Mother—Mother Nature—is essentially ancient pantheism. It is now returning even in "Christian lands", in all its demonic power. When combined with the pervasive controls made possible by modern computerized systems technology, the global goals of evolutionary humanism seem very imminent indeed. Jeremy Rifkin (in his 1983 book, *Algeny*) considers them to be inevitable.

We no longer feel ourselves to be guests in someone else's home and therefore obliged to make our behavior conform with a set of preexisting cosmic rules. It is our creation now. We make the rules. We establish the parameters of reality. We create the world, and because we do, we no longer feel beholden to outside forces. We no longer have to justify our behavior, for we are now the architects of the universe. We are responsible to nothing outside ourselves, for we are the kingdom, the power, and the glory forever and ever.

Rifkin, though certain this is the world's future, is despondent. He closes his book with these words of despair, "Our future is secured. The cosmos wails."

———

And God said, Let the waters bring forth abundantly
The moving creature that hath life,
And the fowl that may fly above the earth
In the open firmament of heaven.

And God created great whales,
And every living creature that moveth,
Which the waters brought forth abundantly, after their kind,
And every winged fowl after his kind:
And God saw that it was good.

—*Genesis 1: 20–21*

———

"Problem? What problem?" *(The Liberal Religious Response)*

The creationist challenge in American education has prompted a variety of responses from scientists, philosophers, and educators who share an evolutionary worldview. Prior to the days of "creation science," which emerged around 1970, antievolutionists cried "blasphemy" rather than "error." Then, a palliative response was credible. Charles Darwin himself had led the way. Anticipating reprobation on religious grounds, Darwin included these passages in the concluding chapter of The Origin of Species.

Charles Darwin　　Although I am fully convinced of the truth of the views given in this volume, I by no means expect to convince experienced naturalists whose minds are stocked with a multitude of facts all viewed, during a long course of years, from a point of view directly opposite to mine. It is so easy to hide our ignorance under such expressions as the "plan of creation," "unity of design," etc., and to think that we give an explanation when we only re-state a fact. Any one whose disposition leads him to attach more weight to unexplained difficulties than to the explanation of a certain number of facts will certainly reject the theory. A few naturalists, endowed with much flexibility of mind, and who have already begun to doubt the immutability of species, may be influenced by this volume; but I look with confidence to the future—to young and rising naturalists, who will be able to view both sides of the question with impartiality. Whoever is led to believe that species are mutable will do good service by conscientiously expressing his conviction; for thus only can the load of prejudice by which this subject is overwhelmed be removed.

Authors of the highest eminence seem to be fully satisfied with the view that each species has been independently created. To my mind it accords better with what we know of the laws impressed on matter by the Creator, that the production and extinction of the past and present inhabitants of the world should have been due to secondary causes, like those determining the birth and death of the individual. When I view all beings not as special creations, but as the lineal descendants of some few beings which lived long before the first bed of the Cambrian system was deposited, they seem to me to become ennobled.

I see no good reason why the views given in this volume should shock the religious feelings of any one. It is satisfactory, as showing how transient such impressions are, to remember that the greatest discovery ever made by man, namely, the law of the attraction of gravity, was also attacked by Leibnitz, "as subversive of natural, and inferentially revealed, religion." A celebrated author and divine has written to me that he has "gradually learnt to see that it is just as noble a conception of the Deity to believe that He created a few original forms capable of self-development into other and needful forms, as to believe that He required a fresh act of creation to supply the voids caused by the action of his laws."

The above paragraphs appeared (unmodified, except that "Cambrian" came to replace "Silurian") in all six editions of The Origin of Species *issued during the lifetime of Charles Darwin. Darwin thus consistently presented himself as an evolutionary theist. In his autobiography, however, which he began composing five years after making changes for the sixth edition of* The Origin, *Darwin confesses the full magnitude of his apostasy. Excerpts here chart the course of his questioning—first, of the verity of the Christian story, then of the existence of a personal God, and finally Darwin's loss of faith in even a far-removed and impersonal God as first cause of the universe.*

Whilst on board the *Beagle* I was quite orthodox, and I remember being heartily laughed at by several of the officers (though themselves orthodox) for quoting the Bible as an unanswerable authority on some point of morality. I suppose it was the novelty of the argument that amused them. But I had gradually come, by this time, to see that the Old Testament from its manifestly false history of the world, with the Tower of Babel, the rainbow as a sign, etc., etc., and from its attributing to God the feelings of a revengeful tyrant, was no more to be trusted than the sacred books of the Hindoos, or the beliefs of any barbarian.

I gradually came to disbelieve in Christianity as a divine revelation. The fact that many false religions have spread over large portions of the earth like wild-fire had some weight with me. Beautiful as is the morality of the New Testament, it can hardly be denied that its perfection depends in part on the interpretation which we now put on metaphors and allegories.

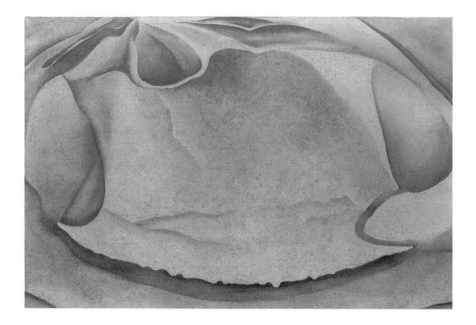

Thus disbelief crept over me at a very slow rate, but was at last complete. The rate was so slow that I felt no distress, and have never since doubted even for a single second that my conclusion was correct. I can indeed hardly see how anyone ought to wish Christianity to be true; for if so the plain language of the text seems to show that the men who do not believe, and this would include my Father, Brother and almost all my best friends, will be everlastingly punished. And this is a damnable doctrine.

Although I did not think much about the existence of a personal God until a considerably later period of my life, I will here give the vague conclusions to which I have been driven. The old argument of design in nature, as given by Paley, which formerly seemed to me so conclusive, fails, now that the law of natural selection has been discovered. We can no longer argue that, for instance, the beautiful hinge of a bivalve shell must have been made by an intelligent being, like the hinge of a door by man. There seems to be no more design in the variability of organic beings and in the action of natural selection, than in the course which the wind blows.

Another source of conviction in the existence of God, connected with the reason and not with the feelings, impresses me as having much more weight. This follows from the extreme difficulty or rather impossibility of conceiving of this immense and wonderful universe, including man with his capacity of looking far backwards and far into futurity, as the result of blind chance or necessity. When thus reflecting I feel compelled to look to a First Cause having an intelligent mind in some degree analogous to that of man; and I deserve to be called a Theist.

This conclusion was strong in my mind about the time, as far as I can remember, when I wrote the *Origin of Species;* and it is since that time that it has very gradually with many fluctuations become weaker. But then arises the doubt—can the mind of man, which has, as I fully believe, been developed from a mind as low as that possessed by the lowest animal, be trusted when it draws such grand conclusions? May not these be the result of the connection between cause and effect which strikes us as a necessary one, but probably depends merely on inherited experience? Nor must we overlook the probability of the constant inculcation in a belief in God on the minds of children producing so strong and perhaps an inherited effect on their brains not yet fully developed, that it would be as difficult for them to throw off their belief in God, as for a monkey to throw off its instinctive fear and hatred of a snake.

I cannot pretend to throw the least light on such abstruse problems. The mystery of the beginning of all things is insoluble by us; and I for one must be content to remain an Agnostic.

A century after Darwin drifted into agnosticism, evolutionary theism as a religion for the scientifically inclined still had (and has) proponents. The geneticist Theodosius Dobzhansky published an essay in 1973 that is cited to this day by peacemakers on the issue of origins. Whereas many other tracts have offered this same solution, Dobzhansky's words carry a special power; he was both distinguished scientist and evolutionary theist.

Dobzhansky wrote at a time when the antievolution laws of the 1920s had all been repealed or invalidated by the courts, and several years before the "balanced treatment" drive got underway. What, then, was his motivation? Simply this: publishers of high school biology texts, looking to minimize conflict and maximize sales, had long downplayed or outright eliminated treatment of evolution. In the 1960s a group of leading biologists who were fed up with commercial inattention to evolution secured funding from the National Science Foundation to develop textbooks that would treat evolution as the central and unifying principle of the discipline. Adoption of these texts touched off a new round of creationist protest, and many schools opted once again for less troublesome fare. Dobzhansky disapproved. Reprinted here courtesy of the National Association of Biology Teachers (Reston, Virginia) are excerpts from the 1973 manifesto, "Nothing in Biology Makes Sense Except in the Light of Evolution," by Theodosius Dobzhansky.

Theodosius Dobzhansky

Seen in the light of evolution, biology is, perhaps, intellectually the most satisfying and inspiring science. Without that light it becomes a pile of sundry facts—some of them interesting or curious but making no meaningful picture as a whole.

This is not to imply that we know everything that can and could be known about biology and about evolution. Any competent biologist is aware of a multitude of problems yet unresolved and of questions yet unanswered. Disagreements and clashes of opinion are rife among biologists, as they should be in a living and growing science. Antievolutionists mistake, or pretend to mistake, these disagreements as indications of dubiousness of the entire doctrine of evolution. Their favorite sport is stringing together quotations, carefully and sometimes expertly taken out of context, to show that nothing is really established or agreed upon among evolutionists. Some of my colleagues and myself have been amused and amazed to read ourselves quoted in a way showing that we are really antievolutionists under the skin.

Let me try to make crystal clear what is established beyond reasonable doubt, and what needs further study, about evolution. Evolution as a process that has always gone on in the history of the earth can be doubted only by those who are ignorant of the evidence or are resistant to evidence, owing to emotional blocks or to plain bigotry. By contrast, the mechanisms that bring evolution about certainly need study and clarification. There are no alternatives to evolution as history that can withstand critical examination. Yet we are constantly learning new and important facts about evolutionary mechanisms.

Does the evolutionary doctrine clash with religious faith? It does not. It is a blunder to mistake the Holy Scriptures for elementary textbooks of astronomy, geology, biology, and anthropology. Only if symbols are construed to mean what they are not intended to mean can there arise imaginary, insoluble conflicts. The blunder leads to blasphemy: the Creator is accused of systematic deceitfulness.

It is wrong to hold creation and evolution as mutually exclusive alternatives. I am a creationist *and* an evolutionist. Evolution is God's, or Nature's, method of Creation.

One of the great thinkers of our age, Pierre Teilhard de Chardin, wrote the following: "Is evolution a theory, a system, or a hypothesis? It is much more—it is a general postulate to which all theories, all hypotheses, all systems must henceforward bow and which they must satisfy in order to be thinkable and true. Evolution is a light which illuminates all facts, a trajectory which all lines of thought must follow." Of course, some scientists, as well as some philosophers and theologians, disagree with some parts of Teilhard's teachings; the acceptance of his world view falls short of universal. But there is no doubt at all that Teilhard was a truly and deeply religious man and that Christianity was the cornerstone of his world view. Moreover, in his world view science and faith were not segregated in watertight compartments, as they are with so many people. They were harmoniously fitting parts of his world view. Teilhard was a creationist, but one who understood that the Creation is realized in this world by means of evolution.

———

How elemental, the way you put it:
In the beginning

was the velocity of light.
But it means that God will never
be the same, now he's plummeted a peg
and had his secrets ogled.

Where does that leave us?
Opting to face God or to face the Truth?
Or to face a God whose Truths are on tap,
some Holy Victim squatting on a pulsar?

Your gospel is no different
from Galileo's, only bolder:
you want to drive a stake
under the nail of the universe
and draw God out
like a soft-shelled crab.

—*Diane Ackerman*

———

"Good fences make good neighbors" (The Separatist Response)

*Pope John Paul II in 1981 stated in no uncertain terms that, in the view
of the Catholic church, biblical fundamentalists were guilty of trespass.*

Pope John Paul II   The Bible itself speaks to us of the origin of the universe and its make-
up, not in order to provide us with a scientific treatise but in order to
state the correct relationships of man with God and with the universe.
Any other teaching about the origin and make-up of the universe is alien
to the intentions of the Bible, which does not wish to teach how the
heavens were made but how one goes to heaven.

*By 1988, the pope was concerned less with the transgressions of biblical
fundamentalists and more with the opportunities for dialogue between
science and religion. In a statement commemorating the three hundredth
anniversary of Newton's* Principia, *Pope John Paul II called for an interaction between science and religion that would steer clear of the extremes
of isolation on the one hand and predatory extension on the other. The
pope's statement appears in* John Paul on Science and Religion: Reflections on the New View from Rome, *edited by Robert J. Russell, William
R. Stoeger, and George V. Coyne. Portions are reprinted here courtesy of
the Vatican Observatory Foundation.*

Turning to the relationship between religion and science, there has been a
definite, though still fragile and provisional, movement towards a new
and more nuanced interchange. We have begun to talk to one another on
deeper levels than before, and with greater openness towards one another's perspectives. We have begun to search together for a more thorough
understanding of one another's disciplines, with their competencies and
their limitations, and especially for areas of common ground.

By encouraging openness between the Church and the scientific communities, we are not envisioning a disciplinary unity between theology and science like that which exists within a given scientific field or within theology proper. As dialogue and common searching continue, there will be growth towards mutual understanding and a gradual uncovering of common concerns which will provide the basis for further research and discussion. Exactly what form that will take must be left to the future. What is important is that the dialogue should continue and grow in depth and scope. In the process we must overcome every regressive tendency to a unilateral reductionism, to fear, and to self-imposed isolation. What is critically important is that each discipline should continue to enrich, nourish and challenge the other to be more fully what it can be and to contribute to our vision of who we are and who we are becoming.

We might ask whether or not we are ready for this crucial endeavor. Is the community of world religions, including the Church, ready to enter into a more thorough-going dialogue with the scientific community, a dialogue in which the integrity of both religion and science is supported and the advance of each is fostered? Is the scientific community now prepared to open itself to Christianity, and indeed to all the great world religions, working with us all to build a culture that is more humane and in that way more divine? Do we dare to risk the honesty and the courage that this task demands? We must ask ourselves whether both science and religion will contribute to the integration of human culture or to its fragmentation. It is a single choice and it confronts us all.

For a simple neutrality is no longer acceptable. If they are to grow and mature, peoples cannot continue to live in separate compartments, pursuing totally divergent interests from which they evaluate and judge their world. A divided community fosters a fragmented vision of the world; a community of interchange encourages its members to expand their partial perspectives and form a new unified vision.

Yet the unity that we seek is not identity. The Church does not propose that science should become religion or religion science. On the contrary, unity always presupposes the diversity and the integrity of its elements. Each of these members should become not less itself but more itself in a dynamic interchange, for a unity in which one of the elements is reduced to the other is destructive, false in its promises of harmony, and ruinous of the integrity of its components. We are asked to become one. We are not asked to become each other.

To be more specific, both religion and science must preserve their autonomy and their distinctiveness. Religion is not founded on science nor is science an extension of religion. Each should possess its own principles, its pattern of procedures, its diversities of interpretation and its own conclusions. Christianity possesses the source of its justification within itself and does not expect science to constitute its primary apologetic. Science must bear witness to its own worth. While each can and should support the other as distinct dimensions of a common human culture, neither ought to assume that it forms a necessary premise for the other. The unprecedented opportunity we have today is for a common interactive relationship in which each discipline retains its integrity and yet is radically open to the discoveries and insights of the other.

Science develops best when its concepts and conclusions are integrated into the broader human culture and its concerns for ultimate meaning and value. Scientists cannot, therefore, hold themselves entirely aloof from the sorts of issues dealt with by philosophers and theologians. By devoting to these issues something of the energy and care they give to their research in science, they can help others realize more fully the human potentialities of their discoveries. They can also come to appreciate for themselves that these discoveries cannot be a genuine substitute for knowledge of the truly ultimate. Science can purify religion from error and superstition; religion can purify science from idolatry and false absolutes. Each can draw the other into a wider world, a world in which both can flourish.

The plea of Pope John Paul II for a mutually supportive and sophisticated dialogue between science and religion has not convinced Protestant fundamentalists to fold up their tents and go home. Creationists still protest the teaching in public schools of evolution (or, rather, the "unbalanced" teaching of origins based on an evolutionary framework alone). A starkly separatist approach, then, has been promoted by some scientists and educators as the best path for preserving the integrity of school science. This approach hails from the very roots of America's constitutional separation of church and state. In 1784 James Madison (then a Virginia legislator who opposed Patrick Henry's bill to tax individuals for the support of each taxpayer's choice of a Christian church) proclaimed, "True religion does not need the help of the state and a just government does not need the help of a church." In Abusing Science, *a 1982 book that takes on the creationists, philosopher of science Philip Kitcher advocates a staunchly separatist resolution to the conflict.*

Philip Kitcher   Evolutionary theorists and educators do not fear the evidence. There is no doubt that a fair presentation of the evidence, and a careful review of the arguments, will support evolutionary theory and unmask "scientific" Creationism for what it is. What is in doubt is the possibility of a fair and complete presentation of the issues in the context of the high school classroom. Even a gifted teacher would not be able to expound enough of the scientific background to make it clear that all the salvos [of the creationists] miss their mark. (Would it even be permissible for the teacher to expose Creationist distortions?) What Creationists really propose is a situation in which people without scientific training—fourteen-year-old students, for example—are asked to decide a complex issue on partial evidence.

Anthropologist Ashley Montagu similarly views creationism as an inappropriate subject for the science classroom. This excerpt is drawn from a collection of essays he assembled for his 1984 book, Science and Creationism.

Ashley Montagu   All that Senator Keith *(a Louisiana legislator)* asked was that "creation-science receive balanced treatment" with "evolution-science," that wherever in the schools "evolution-science" is taught, "creation-science" also be taught. This is, of course, the equivalent of requiring that wherever chemistry is taught alchemy should be given equal time, or wherever psychology is taught phrenology be given balanced treatment, or astrology be given equal time with astronomy.

On the matter of religion and science, it needs to be said that there is no real incompatibility between a belief in God and the belief that evolution is the means by which all living things have come into being. What *is* incompatible with science, religion, and civility is the attempt by a narrow fundamentalist sect to impose its particular brand of a creation myth as a substitute or alternative to the findings of science, and to insist on having that myth taught as a fact in the schools, as to the exclusion of all other religious teachings.

Niles Eldredge, a paleontologist and evolutionary theorist, is confident of a separatist win in the courts, but he is worried that the victory may be hollow.

Niles Eldredge   The creationists will not win in court on the trumped-up charge that evolution is "secular humanism" and not science. But they *may* win in the more important arena of public opinion if they succeed (as they have to a remarkable degree thus far) in convincing our fellow citizens that science is just another authoritarian belief system, and that Americans, in the traditional sense of "fair play," should be allowed to "hear both sides."

Stephen Jay Gould, also a paleontologist and evolutionary theorist (and codeveloper with Niles Eldredge of "punctuated equilibrium," a controversial twist on neo-Darwinist evolutionary theory), warns that the problem of creationism for educators in the public schools is spilling over into the halls of academe—with unfortunate consequences.

Stephen Jay Gould   I am both angry at and amused by the creationists; but mostly I am deeply sad. I am sad because the practical result of this brouhaha will not be expanded coverage to include creationism (that would also make me sad), but the reduction or excision of evolution from high school curricula. Evolution is one of the half dozen "great ideas" developed by science. It speaks to the profound issues of genealogy that fascinate all of us—the "Roots" phenomenon writ large. Where did we come from? When did life arise? How did it develop? How are organisms related? It forces us to think, ponder, and wonder. Shall we deprive millions of this

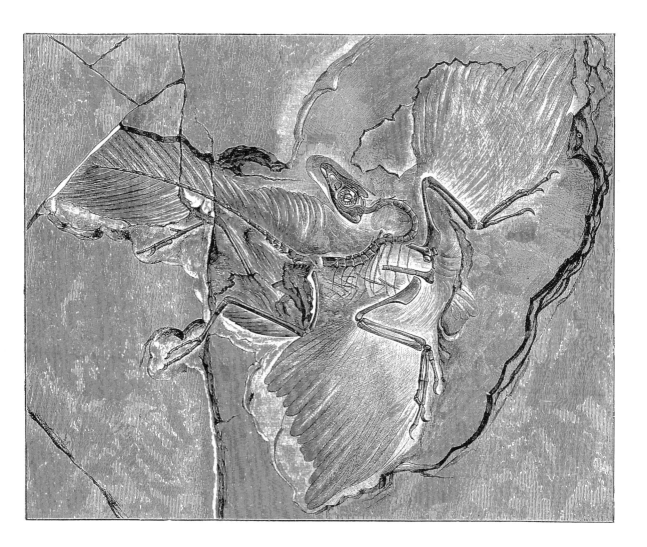

knowledge and once again teach biology as a set of dull and unconnected facts, without the thread that weaves diverse material into a supple unity?

Most of all I am saddened by a trend I am just beginning to discern among my colleagues. I sense that some now wish to mute the healthy debate about theory that has brought new life to evolutionary biology. It provides grist for creationist mills, they say, even if only by distortion. Perhaps we should lay low and rally round the flag of strict Darwinism, at least for the moment—a kind of old-time religion on our part. But we should borrow another metaphor and recognize that we too have to tread a straight and narrow path, surrounded by roads to perdition. For if we ever begin to suppress our search to understand nature, to quench our own intellectual excitement in a misguided effort to present a united front where it does not and should not exist, then we are truly lost.

Karl Popper, the philosopher of science who in chapter 5 spoke to the role of behavioral initiative in evolution, chastises those who stray from the separatist fold.

Karl Popper

It is almost unbelievable how much the atmosphere changed as a consequence of the publication, in 1859, of *The Origin of Species*. Our whole outlook, our picture of the universe, has changed, as never before. The Darwinian revolution is still proceeding. But now we are also in the midst of a counter-revolution, a strong reaction against science and against rationality. I feel that it is necessary to take sides in this issue.

My position, very briefly, is this. I am on the side of science and of rationality, but I am against those exaggerated claims for science that have sometimes been, rightly, denounced as "scientism". I am on the side of the search for truth, and of intellectual daring in the search for truth; but I am against intellectual arrogance, and especially against the misconceived claim that we have the truth in our pockets, or that we can approach certainty. It is important to realize that science does not make assertions about ultimate questions—about the riddles of existence, or about man's task in this world.

John C. Greene is a historian of science who, in the previous chapter, criticized Teilhard de Chardin and Julian Huxley for extending evolutionary biology into the realm of meaning. In his 1981 book, Science, Ideology, and World View, *Greene echoes Karl Popper.*

John C. Greene

Every great scientific synthesis stimulates efforts to view the whole of reality in its terms, and Darwin's theory of natural selection was no exception. But the views of reality that originate in this way are not themselves scientific, nor are they subject to scientific verification. They attempt to make sense not only of the facts "out there," held at arm's length by the observer, but also of the facts "in here," facts such as our awareness of our own act of existence, our appreciation of beauty, our sense of moral accountability, our communion with the source of being. Facts of the latter kind lie close to the heart of reality, but they do not lend themselves to scientific formulation. Attempts to explain them scientifically end by explaining them away. But science itself then becomes unintelligible.

To ignore the differences between science, philosophy, and religion and roll them all into one evolutionary gospel claiming to disclose the meaning of existence is as dangerous to science as it is to philosophy and religion. If scientists aspire to be prophets and preachers, they cannot expect society to grant them the relative autonomy they have enjoyed in Western culture in recent centuries. The current misguided campaign to require the teaching of "creationist biology" alongside evolutionary biology is sufficient evidence of that. The hard-won ideal of disinterested inquiry guided by insight and logic but rigorously controlled by generally accepted methods of empirical testing is too precious an acquisition of the human spirit to be sacrificed to grandiose but delusory and self-destructive dreams of an omnicompetent science of nature-history, society, and human duty and destiny.

As a student of the history of ideas, I am convinced that science, ideology, and world view will forever be intertwined and interacting. As a citizen concerned for the welfare of science and of mankind generally, however, I cannot but hope that scientists will recognize where science ends and other things begin.

———

But as for certain truth, no man has known it,
Nor will he know it; neither of the gods,
Nor yet of all the things of which I speak.
And even if by chance he were to utter
The final truth, he would himself not know it;
For all is but a woven web of guesses.

—*Zenophanes*

———

"Take your religion and . . . !" (The Hardliner Response)

The power of the separatist argument resides in an almost catechismal adherence by evolution theorists and other prominent scientists to the separatist doctrine. Diplomacy is key. Religion as a whole and the creationist faith in particular, if mentioned at all, are treated with nothing short of respect. But some scientists will not buy into what is to them a mere charade of kindness. Francis Bacon is the forerunner of today's provocateurs. In his 1620 book, Novum Organum, *Bacon pulls no punches.*

Francis Bacon

Some of the moderns have with extreme levity indulged so far as to attempt to found a system of natural philosophy on the first chapter of Genesis, the book of Job, and other parts of Scriptures. From this unwholesome mixture of things human and divine there arises not only a fantastic philosophy but also a heretical religion. Very meet it is therefore that we be sober-minded and give to faith that only which is faith's.

Philosopher of science Mary Midgley expresses a similar viewpoint in her 1985 book, Evolution as Religion.

Mary Midgley

In what sense can two such abstract entities as science and religion (or morality) be said to clash? Mere accidental personal feuds between their followers are not enough to justify this language. They can surely only clash where they compete, where they represent rival attempts to perform the same function. How far, if at all, can science and religion do this?

There is, of course, a well-known set of cases where they seem to do it, namely, where religion is invoked against science on a point of empirical fact. The literal acceptance of archaic Biblical ideas on cosmology is an obvious case. Creationists who attempt this are taking on a scientific task, as indeed they now recognize by their preference for talk of 'creation science'. But their reasons for undertaking it flow not from religion as such, nor even from Christianity, but from their own peculiar conception of the Bible as literally true and divinely dictated.

Other Christians object to this view strongly, on the obvious grounds that it is needless, and moreover that the Bible, in spite of its grandeur, contains many things which conflict not just with science, but with morality, with history, with common sense, or with each other. If there were a god who had dictated the whole of it, he would certainly not be one we ought to worship. Biblical writers seem, then, to have been as fallible and imperfect as other human beings, and moreover to have used—as would naturally be expected—a mythical and metaphorical way of writing where that was suitable, instead of making the quite irrelevant attempt to be modern physical scientists.

The central objections to fundamentalist literalism are religious, moral and historical ones. If they are right, this is a case where 'religion' does not clash with science unless something has gone wrong with it already on its own terms. The religion which does clash with science has left its own sphere, for bad reasons, to intrude on a scientific one. It is bad religion.

George Gaylord Simpson, in his 1964 book, This View of Life: The World of an Evolutionist, *gives no ground—not even religious ground— to exponents of creationism. (Simpson, you may recall, was foil to Julian Huxley in chapter 2, which presented the debate about evolutionary progress.)*

George Gaylord Simpson

The conflict between science and religion has a single and simple cause. It is the designation as religiously canonical of any conception of the material world open to scientific investigation. That is a basis for conflict even when religion and science happen to agree as to the material facts. The religious canon demands absolute acceptance not subject to test or revision. Science necessarily rejects certainty and predicates acceptance on objective testing and the possibility of continual revision. As a matter of fact, most of the dogmatic religions have exhibited a perverse talent for taking the wrong side on the most important concepts of the material universe, from the structure of the solar system to the origin of man. The result has been constant turmoil for many centuries, and the turmoil will continue as long as religious canons prejudge scientific questions.

If a sect does officially insist that its structure of belief demands that evolution be false, then no compromise is possible. An honest and competent biology teacher can only conclude that the sect's beliefs are wrong and that its religion is a false one. Whatever else God may be, He is surely consistent with the world of observed phenomena in which we live. A god whose means of creation is not evolution is a false god. It is not the teacher's duty to point this out unnecessarily, but it is certainly his duty not to compromise the point.

Fortunately, the great majority of religious people in America belong to sects that are more flexible on this point, even though the tendency of the average parishioner may be antievolutionary. Here a perfectly honest compromise, or rather a tolerant understanding, is possible. Evolution, per se, is not antireligious any more than the roundness of the earth is antireligious, although it was once held to be so. There are many religious and, in various sects, even highly orthodox evolutionists. There are also atheistic evolutionists, but so are there atheistic bankers, who nevertheless keep honest accounts. The lack of necessity for conflict between evolution and religion is something that can and, when the subject arises, should be pointed out by teachers. The most extreme and bigoted opponents cannot be placated, but there is plenty of common ground for reasonable people on this question.

There are, to be sure, many high schools where evolution is taught without opposition from students or community and even with their enthusiastic support. There are also textbooks that include evolution under its right name and as an established biological fact. Nevertheless, it is certainly true that innumerable students still leave high school without ever having heard of evolution, or having heard of it only in such a way as to leave them unimpressed or antagonistic. Since intermediate education is the proper level for encountering this subject and is for great numbers of people the only place where they are likely to learn anything valid about it, this means that an awareness of evolution is lacking or rejected in large segments of the adult population. Yet for over a century now, evolution has been known to be one of the great and central concepts of science and one fundamental for human orientation in the modern world. There is no other concept of comparable importance and scope that has been so slow in permeating education and in obtaining general popular acceptance.

The rational, that is, the scientific investigation of the universe reveals its marvels as no amount of introspection or revelation could do. We are in ourselves truly fearful and wonderful, and so is a skylark, a buttercup, or indeed a grain of sand. No poet or seer has ever contemplated wonders as deep as those revealed to the scientist. Few can be so dull as not to react to our material knowledge of this world with a sense of awe that merits designation as religious.

Bacon, Midgley, and Simpson all sound like peacemakers compared to the provocateurs who follow. William Provine is both a respected historian of evolutionary biology and an outspoken hardliner on the issue of creationism. This excerpt is drawn from his essay "Progress in Evolution and Meaning in Life," which appeared in Evolutionary Progress *(Matthew Nitecki, editor; copyright 1988 University of Chicago Press). It is reprinted here by permission of the author and the publisher.*

William Provine

According to the unlikely coalition of scientists, jurists, educators, theologians, and religious leaders, the conflict between science and religion exists only in the naive, literalist minds of the creationists. Scientists testify that creation science is stupid and so is the literalist religion of the creationists. The federal courts and even now the Supreme Court say the same things. So do the religious leaders. Nearly half of the American public is being called stupid or at best misguided. I agree that evolution has occurred and that creationists are wrong about that. But they are right that there is a conflict between science and religion, not only their religions.

The conflict is fundamental and goes much deeper than modern liberal theologians, religious leaders, and scientists are willing to admit. Most contemporary scientists, the majority of them by far, are atheists or something very close to that. And among evolutionary biologists, I would challenge the reader to name the prominent scientists who are "devoutly religious." I am skeptical that one could get beyond the fingers of one hand. Indeed, I would be interested to learn of a single one. Osborn, Lack, Dobzhansky, and Fisher are dead, and no generation of compatibilists among prominent evolutionary biologists has replaced them.

I suspect there is a lot of intellectual dishonesty on this issue. Consider the following fantasy: the National Academy of Sciences publishes a position paper on science and religion stating that modern science leads directly to atheism. What would happen to its funding? To any federal funding of science? Every member of the Congress of the United States of America, even the two current members who are unaffiliated with any organized religion, professes to be deeply religious. I suspect that scientific leaders tread very warily on the issue of the religious implications of science for fear of jeopardizing the funding for scientific research. And I think that many scientists feel some sympathy with the need for moral education and recognize the role that religion plays in this endeavor. These rationalizations are politic but intellectually dishonest.

Many theologians have reacted to the rise of modern science by retreating from traditional conceptions of God and its presence in the world, calling this a more sophisticated view. God used to be all around us earlier in our cultural history. It used to perform miracles. It used to guide its people. People could detect God's presence all the time; but times have changed. God is more remote today. In fact, one cannot rationally discover anything that God does in the world anymore. A widespread theological view now exists saying that God started off the world, props it up and works through laws of nature, very subtly, so subtly that its action is undetectable. But that kind of God is effectively no different to my mind than atheism. To anyone who adopts this view I say, "Great, we're in the same camp; now where do we get our morals if the universe just goes grinding on as it does?" This kind of God does nothing outside of the laws of nature, gives us no immortality, no foundation for morals, or any of the things that we want from a God and from religion.

I can see no cosmic or ultimate meaning in life. Julian Huxley's dream that evolution could provide the basis for ethics and give meaning to life is just that—a dream. The universe cares nothing for us and will probably either continue to expand and cool, leading to an extermination of all living creatures in the universe, or the universe will cease to expand and will begin closure, which will result in everything crashing together in an unbelievably small space, thus obliterating all life. Even if the universe is in some sort of equilibrium and both scenaria above are wrong, the universe has so far exhibited no care for humans and gives no rational hope of future caring.

Humans are as nothing even in the evolutionary process on earth, and only a few individuals are remembered for as many as ten generations. There is no ultimate meaning for humans. But certainly humans can lead meaningful lives. My own life is filled with meaning. I am married to an intellectually talented and beautiful woman, have two great sons, live on a 150-acre farm with pond, river, wild turkeys, and lots of old but good farm machinery; I teach at a fine university with excellent students and have many wonderful friends. But I will die and soon be forgotten.

There is no better way to close this section on the hardliner response than with the wit of Richard Dawkins, one of today's top evolutionary theorists, whose depiction of "evolutionary arms races" appeared in chapter 3.

Richard Dawkins　　Nearly all peoples have developed their own creation myth, and the Genesis story is just the one that happened to have been adopted by one particular tribe of Middle Eastern herders. It has no more special status than the belief of a particular West African tribe that the world was created from the excrement of ants.

———

How many turn back toward dreams and magic,
　　how many children
Run home to Mother Church, Father State,
To find in their arms the delicious warmth and folding
　　of souls.
The age weakens and settles home toward old ways.
An age of renascent faith: Christ said, Marx wrote,
　　Hitler says,
And though it seems absurd we believe

Sad children, yes. It is lonely to be adult, you need
   a father.
With a little practice you'll believe anything.

—*Robinson Jeffers*

———

"Let's see how we can learn from this" (The Conciliatory Response)

As the creationist lobby has dropped its old antievolutionism and now urges "balanced treatment" in education, so have some evolutionists devised a more conciliatory strategy. Tossing aside separatism, these conciliators see a role for discussion of creationism in public school classrooms. Their reasons, however, are diverse and sometimes conflicting. A quotation of Alfred North Whitehead is nevertheless an apt characterization of the conciliatory view in the main: "A clash of doctrines is not a disaster—it is an opportunity." This overview of the conciliatory response to creationism begins with excerpts from an essay by physicist and philosopher Martin Eger. "A Tale of Two Controversies: Dissonance in the Theory and Practice of Rationality" appeared in the September 1988 issue of Zygon: Journal of Religion and Science.

Martin Eger

On every subject on which difference of opinion is possible, the truth depends on a balance to be struck between two sets of conflicting reasons. Even in natural philosophy, there is always some other explanation possible of the same facts; some geocentric theory instead of heliocentric, some phlogiston instead of oxygen; and it has to be shown why that other theory cannot be the true one: and until this is shown, and until we know how it is shown, we do not understand the grounds of our opinion.

—*John Stuart Mill,* On Liberty, 1859

Mill, writing with a judgment whetted by philosophy of science, reached this position more than 300 years after Copernicus, some 250 years after Galileo's discoveries. Today, following his logic, we may well ask whether "keeping creationism out of the schools" is not one reason why biology teachers are often incompetent to answer the classic challenges to evolution—never having faced these challenges themselves as did Darwin.

Anti-creationist writers routinely raise the fear of a scientific dark age if evolution is not taught within the framework of a totally dominant paradigm, if instead it is encumbered by distracting, useless comparisons with a theory long outworn. In this type of argument, all interest lies within the context of application. For it is problems that occupy center-stage, where students are seen as potential problem-solvers.

By contrast, those who take seriously the creationist argument, or merely its critical side in relation to a magnified evolutionary vision, are oriented almost exclusively to issues within the context of education. They worry about what evolution says, or implies, or may imply to young minds, concerning the roots of human life. They think this important because beliefs about origin do contribute to our image of self, which, in turn, affects our human interaction. And they have strong misgivings about a "scientific world view" so pervasive and well articulated that it expels from consciousness all frameworks not based on an imperative of problem-solving.

Granted that to recognize the role of different contexts in these conflicts is hardly to resolve the conflicts; nor, in the context of education, does creationism somehow become a better theory. But it is also clear that the conventional argument for treating science differently fails. It fails because the difference in object domains and the kinds of questions asked is not sufficient reason for the greatly diminished role of the critical stance in science study as compared with morals—provided the comparison is in the context of education. How can the scientific, skeptical, critical attitude be an example for moral education and other studies if, for the sake of expediency or professional need, science courses themselves eschew it? And how can we require the next generation to accept on authority our scientific tradition if all other traditions are ignored or cheerfully dismembered?

The book On Liberty *by John Stuart Mill, from which Martin Eger extracted an excerpt, can be mined for many more quotations that bolster arguments by today's conciliators. Here are a few of the best.*

John Stuart Mill

However unwillingly a person who has a strong opinion may admit the possibility that his opinion may be false, he ought to be moved by the consideration that however true it may be, if it is not fully, frequently, and fearlessly discussed, it will be held as a dead dogma, not a living truth. . . .

On any subject no one's opinions deserve the name of knowledge, except so far as he has either had forced upon him by others, or gone through himself, the same mental process which would have been required of him in carrying on an active controversy with opponents. That, therefore, which when absent, it is so indispensable, but so difficult, to create, how worse than absurd it is to forgo, when spontaneously offering itself! If there are any persons who contest a received opinion, let us thank them for it, open our minds to listen to them, and rejoice that there is someone to do for us what we otherwise ought, if we have any regard for either the certainty or the vitality of our convictions, to do with much greater labour for ourselves. . . .

Both teachers and learners go to sleep at their post as soon as there is no enemy in the field.

John Stuart Mill's book was published in 1859—the same year as The Origin of Species. *In 1894 Thomas Henry Huxley, nicknamed "Darwin's bulldog" for his feisty defense of Darwin's grand idea, published* Darwiniana. *I was surprised to discover there an argument that echoes Mill.*

Thomas Huxley	History warns us that it is the customary fate of new truths to begin as heresies and to end as superstitions; and, as matters now stand, it is hardly rash to anticipate that, in another twenty years, the new generation, educated under the influences of the present day, will be in danger of accepting the main doctrines of the *Origin of Species* with as little reflection, and it may be with as little justification, as so many of our contemporaries, twenty years ago, rejected them.

Back to today's debate, Dorothy B. Rosenthal (formerly coordinator of science for the public schools of Henrietta, New York, and now an associate professor of science eduction at California State University) advocates a conciliatory response.

Dorothy B. Rosenthal	Evolution, no matter how central it is to biology and no matter how much support it has among biologists, is still a matter of controversy among certain segments of the population; hence it is a social issue. Furthermore, the teaching of evolution in public schools has been even more controversial; hence evolution education is a social issue. As to the place of social issues in the curriculum, I believe that in addition to teaching evolution, for example, in a scientifically sound manner, we owe it to students to inform them that there is a controversy in society about evolution and to deal honestly with that controversy. I do not view this, in any way, as falling for "the Creationists' recent ploys," but as strengthening our students' understanding of science as an ongoing, open system and as a human endeavor which interacts with the social environment. Science education will be improved by relating science to the many important issues of the day which arise from the growth of science, its increasing impact on society and the interrelationships between science and society.

Kenneth W. Kemp, a professor of philosophy at the United States Air Force Academy in Colorado Springs, argues for a conciliatory response.

Kenneth W. Kemp	It is not fair to students who may have heard something about creation science to dismiss it summarily without a word as to what is wrong with it. Surely, there is something to be said about why it is bad. And understanding why scientists reject inadequate theories is part of what it is to be educated about science.

Similarly, Michael Scriven puts the issue into the perspective of student lives outside the classroom.

Michael Scriven	It is sometimes said that the problem with critical discussion of sensitive matters in the school is that the students are not mature enough to handle them. The first fallacy with this argument is the idea that the students are not already facing major decisions demanding every resource of the critical mind. The second fallacy is the assumption that the students will

mentally mature when they leave school in some mysterious way that will offset ignorance.

Richard D. Alexander, a prominent evolutionary theorist at the University of Michigan, has no truck with diplomacy when it comes to creationism. Nevertheless, he argues for an educational climate devoted to free inquiry.

Richard D. Alexander

When a creationist, Darwinist, Marxist, or supporter of any other theory defends his or her views publicly, he or she does everyone a service. But when anyone attempts to establish laws or rules requiring that certain theories be taught or not be taught, he or she invites us to take a step toward totalitarianism. Whether a law is to prevent the teaching of a theory or to require it is immaterial.

No laws were ever passed saying that evolution had to be taught in biology courses. The prestige of evolutionary theory has been built by its impact on the thousands of biologists who have learned its power and usefulness in the study of living things. No laws need to be passed for creationists to do the same thing.

When creation theorists strive to introduce creation into the classroom as an alternative biological theory to evolution they must recognize that they are required to give creation the status of a falsifiable idea—that is, an idea that loses any special exemption from scrutiny, that is accepted as conceivably being false, and that must be continually tested until the question is settled. A science classroom is not the place for an idea that is revered as holy.

The greatest threat to society and to our children is not whether students are exposed to wrong ideas—after all, many high school biology students are legally adults with voting privileges, and all high school biology students have already been exposed to many wrong ideas. What is important is whether each has been taught how and given the freedom to test new ideas, evaluate them, and respond appropriately. The question of whether evolution or creation or both are mentioned, supported, or taught in any or all of the schools is trivial by comparison. As long as biology teachers conduct their courses in the spirit of free inquiry, open debate, and self-correcting searches for predictive theories and repeatable results, no parent need fear that his or her children are being subjected to anything but the best kind of preparation for life in the technologically complex and socially demanding society in which we live.

———

We want the suns and moons
of silver in ourselves, not only
counted coins in a cup. The whole idea
of love was not to fall. And neither was
the whole idea of god. We put him well

above ourselves because we meant,
in time, to measure up.

—*Heather McHugh*

———

The Supreme Court Rules on Balanced Treatment

*In 1987 the United States Supreme Court overturned the Balanced Treat-
ment for Creation-Science and Evolution-Science Act, enacted by Louisiana in 1982. Justice Brennan wrote the majority opinion, in which
Justices Marshall, Blackmun, Powell, Stevens, O'Connor, and White either joined or wrote separate but concurring opinions. Justice Scalia filed
a dissenting opinion, in which Chief Justice Rehnquist joined. The now-
invalidated Louisiana act reads:*

Public schools within [the] state shall give balanced treatment to creation-
science and to evolution-science. Balanced treatment of these two models
shall be given in classroom lectures taken as a whole for each course, in
textbook materials taken as a whole for each course, in library materials
taken as a whole for the sciences and taken as a whole for the humanities, and in other educational programs in public schools, to the extent
that such lectures, textbooks, library materials, or educational programs
deal in any way with the subject of the origin of man, life, the earth, or
the universe. When creation or evolution is taught, each shall be taught
as a theory, rather than as a proven fact.

*Following are excerpts from the majority and the minority opinions.
Paragraphs may appear out of order, and there are, of course, many deletions. (Consult the concordance.) These excerpts have been chosen, however, to preserve the heart of the arguments, the stylistic clarity, and the
evident concern for individuals and society expressed by the justices—
both those who spoke for the Court and those who dissented. First, the
majority:*

U.S. Supreme Court The question for decision is whether Louisiana's "Balanced Treatment for
(Majority) Creation-Science and Evolution-Science in Public School Instruction" Act
(Creationism Act) is facially invalid as violative of the Establishment
Clause of the First Amendment.

The Establishment Clause forbids the enactment of any law "respecting
an establishment of religion." The Court has applied a three-pronged test
to determine whether legislation comports with the Establishment Clause.
First, the legislature must have adopted the law with a secular purpose.
Second, the statute's principal or primary effect must be one that neither
advances nor inhibits religion. Third, the statute must not result in an
excessive entanglement of government with religion. State action violates
the Establishment Clause if it fails to satisfy any of these prongs.

The Court has been particularly vigilant in monitoring compliance with the Establishment Clause in elementary and secondary schools. Families entrust public schools with the education of their children, but condition their trust on the understanding that the classroom will not purposely be used to advance religious views that may conflict with the private beliefs of the student and his or her family. Students in such institutions are impressionable and their attendance is involuntary. The State exerts great authority and coercive power through mandatory attendance requirements, and because of the students' emulation of teachers as role models and the children's susceptibility to peer pressure. Furthermore, "the public school is at once the symbol of our democracy and the most pervasive means for promoting our common destiny. In no activity of the State is it more vital to keep out divisive forces than in its schools." *(I have eliminated all citations; an unattributed quotation signals a statement drawn from an earlier decision of the Supreme Court.)*

There can be no legitimate state interest in protecting particular religions from scientific views "distasteful to them." We do not imply that a legislature could never require that scientific critiques of prevailing scientific theories be taught. Indeed, the Court acknowledged in *Stone v. Graham* that its decision forbidding the posting of the Ten Commandments did not mean that no use could ever be made of the Ten Commandments, or that the Ten Commandments played an exclusively religious role in the history of Western Civilization. In a similar way, teaching a variety of scientific theories about the origins of humankind to schoolchildren might be validly done with the clear secular intent of enhancing the effectiveness of science instruction. But because the primary purpose of the Creationism Act is to endorse a particular religious doctrine, the Act furthers religion in violation of the Establishment Clause.

The Creationism Act forbids the teaching of the theory of evolution in public schools unless accompanied by instruction in "creation science." While the Court is normally deferential to a State's articulation of a secular purpose, it is required that the statement of such purpose be sincere and not a sham. It is clear from the legislative history that the purpose of the legislative sponsor, Senator Bill Keith, was to narrow the science curriculum. During the legislative hearings Senator Keith stated: "My preference would be that neither [creationism nor evolution] be taught." Such a ban on teaching does not promote—indeed, it undermines—the provision of a comprehensive scientific education.

It is equally clear that requiring schools to teach creation science with evolution does not advance academic freedom. The Act does not grant teachers a flexibility that they did not already possess to supplant the present science curriculum with the presentation of theories, besides evolution, about the origin of life. Indeed, the Court of Appeals *(the court that first overturned the law in question, which prompted an appeal to the Supreme Court)* found that no law prohibited Louisiana public schoolteachers from teaching any scientific theory. Thus we agree with the Court of Appeals' conclusion that the Act does not serve to protect academic freedom, but has the distinctly different purpose of discrediting "evolution by counterbalancing its teaching at every turn with the teaching of creation science."

The Louisiana Creationism Act advances a religious doctrine by requiring either the banishment of the theory of evolution from public school classrooms or the presentation of a religious viewpoint that rejects evolution in its entirety. The Act violates the Establishment Clause of the First Amendment because it seeks to employ the symbolic and financial support of government to achieve a religious purpose. The judgment of the Court of Appeals therefore is affirmed.

———

From earliest schooldays
children are shown
how small the earth is,
how short man's life
is a basketball, the earth
is a cherrypit in the next county.
If the planet is an hour old,
man has only been here
the wink of an eye. And that eye
would not see you, seven years old,
standing rapt in the museum
awed by dinosaur bones. It notices
only seas filling in, mountains
popping up. If the sun
is an orange, this poem
is bigger than the whole world.

—*Laura Fargas*

———

The Dissent

The minority opinion of the U.S. Supreme Court was written by Justice Scalia and joined by Chief Justice Rehnquist. Note that their dissent extends even to the choice of abbreviated title for the Louisiana law under consideration. The majority refers to it as the Creationism Act; the minority calls it the Balanced Treatment Act. Here are key excerpts from the minority opinion:

U.S. Supreme Court (Minority)   Even if I agreed with the questionable premise that legislation can be invalidated under the Establishment Clause on the basis of its motivation alone, without regard to its effects, I would still find no justification for today's decision. The Louisiana legislators who passed the "Balanced Treatment for Creation-Science and Evolution-Science Act" (Balanced Treatment Act), each of whom had sworn to support the Constitution, were well aware of the potential Establishment Clause problems and considered that aspect of the legislation with great care. After seven hearings and several months of study, resulting in substantial revision of the original proposal, they approved the Act overwhelmingly and specifically articulated the secular purpose they meant it to serve. Although the record contains abundant evidence of the sincerity of that purpose (the only issue pertinent to this case), the Court today holds, essentially on the basis of "its visceral knowledge regarding what must have motivated the legislators" *(this quotation is drawn from the dissenting statement in the court of appeals ruling),* that the members of the Louisiana Legislature knowingly violated their oaths and then lied about it. I dissent.

Our cases in no way imply that the Establishment Clause forbids legislators merely to act upon their religious convictions. We surely would not strike down a law providing money to feed the hungry or shelter the homeless if it could be demonstrated that, but for the religious beliefs of the legislators, the funds would not have been approved. Also, political activism by the religiously motivated is part of our heritage. Notwithstanding the majority's implication to the contrary, we do not presume that the sole purpose of a law is to advance religion merely because it was supported strongly by organized religions or by adherents of particular faiths. To do so would deprive religious men and women of their right to participate in the political process. Today's religious activism may give us the Balanced Treatment Act, but yesterday's resulted in the abolition of slavery, and tomorrow's may bring relief for famine victims. Similarly, we will not presume that a law's purpose is to advance religion merely because it "happens to coincide or harmonize with the tenets of some or all religions." Thus, the fact that creation science coincides with the beliefs of certain religions, a fact upon which the majority relies heavily, does not itself justify invalidation of the Act.

The case arrives here in the following posture: The Louisiana Supreme Court has never been given an opportunity to interpret the Balanced Treatment Act, State officials have never attempted to implement it, and it has never been the subject of a full evidentiary hearing. We can only guess at its meaning. We know that it forbids instruction in either "creation-science" or "evolution-science" without instruction in the other, but the parties are sharply divided over what creation science consists of. Appellants insist that it is a collection of educationally valuable scientific data that has been censored from classrooms by an embarrassed scientific establishment. Appellees insist it is not science at all but thinly veiled

religious doctrine. Both interpretations of the intended meaning of the phrase find considerable support in the legislative history.

At least at this stage in the litigation, it is plain to me that we must accept appellants' view of what the statute means. At this point, then, we must assume that the Balanced Treatment Act does not require the presentation of religious doctrine. Had requirements of the Balanced Treatment Act that are not apparent on its face been clarified by an interpretation of the Louisiana Supreme Court, or by the manner of its implementation, the Act might well be found unconstitutional; but the question of its constitutionality cannot rightly be disposed of on the gallop, by impugning the motives of its supporters.

The Louisiana Legislature explicitly set forth its secular purpose ("protecting academic freedom") in the very text of the Act. The Court seeks to evade the force of this expression of purpose by stubbornly misinterpreting it, and then finding that the provisions of the Act do not advance that misinterpreted purpose, thereby showing it to be a sham. The Court first surmises that "academic freedom" means "enhancing the freedom of teachers to teach what they will," even though "academic freedom" in that sense has little scope in the structured elementary and secondary curriculums with which the Act is concerned. Alternatively, the Court suggests that it might mean "maximizing the comprehensiveness of effectiveness of science instruction"—though that is an exceedingly strange interpretation of the words, and one that is refuted on the very face of the statute.

Had the Court devoted to this central question of the meaning of the legislatively expressed purpose a small fraction of the research into legislative history that produced its quotations of religiously motivated statements by individual legislators, it would have discerned quite readily what "academic freedom" meant: students' freedom from indoctrination. The legislature wanted to ensure that students would be free to decide for themselves how life began, based upon a fair and balanced presentation of the scientific evidence. As originally introduced, the "purpose" section of the Balanced Treatment Act read: "This Chapter is enacted for the purposes of protecting academic freedom . . . of students . . . and assisting students in their search for truth." Among the proposed findings of fact contained in the original version of the bill was the following: "Public school instruction in only evolution-science . . . violates the principle of academic freedom because it denies students a choice between scientific models and instead indoctrinates them in evolution science alone."

In sum, even if one concedes, for the sake of argument, that a majority of the Louisiana Legislature voted for the Balanced Treatment Act partly in order to foster (rather than merely eliminate discrimination against) Christian fundamentalist beliefs, our cases establish that that alone would not suffice to invalidate the Act, so long as there was a genuine secular purpose as well. We have, moreover, no adequate basis for disbelieving the secular purpose set forth in the Act itself, or for concluding that it is a sham enacted to conceal the legislators' violation of their oaths of office. I am astonished by the Court's unprecedented readiness to reach such a conclusion.

The people of Louisiana, including those who are Christian fundamentalists, are quite entitled, as a secular matter, to have whatever scientific evidence there may be against evolution presented in their schools, just as Mr. Scopes *(a schoolteacher tried in 1927 for violating a Tennessee law that banned the teaching of evolution)* was entitled to present whatever scientific evidence there was for it. Perhaps what the Louisiana Legislature has done is unconstitutional because there is no such evidence, and the scheme they have established will amount to no more than a presentation of the Book of Genesis. But we cannot say that on the evidence before us in this summary judgment context, which includes ample uncontradicted testimony that "creation science" is a body of scientific knowledge rather than revealed belief.

Because the majority rules, the opinion written by Justice Brennan, not Justice Scalia, has the force of law. But does the ruling ensure a lasting victory for proponents of evolutionary biology in public schools? Edward J. Larson, former counsel for the U.S. House of Representatives Committee on Education and Labor (and who now teaches history and law at the University of Georgia), predicts otherwise. This excerpt is drawn from the 1989 edition of his book, Trial and Error: The American Controversy over Creation and Evolution.

Edward J. Larson   Ultimately, the law could not resolve the controversy because neither side would accept an adverse result as final. Creationists petitioned legislatures, courts, and school boards for relief from evolutionary teaching, while evolutionists sought similar relief from creationist instruction. When either side won a victory, the other one typically redoubled its efforts. Because the creation-evolution controversy remained unresolved in popular opinion, it could not be settled in law.

The controversy over evolutionary teaching is as lively today as ever. [In the majority opinion of the U.S. Supreme Court, which struck down the Louisiana Balanced Treatment Act] Justice Brennan added that "teaching a variety of scientific theories about the origins of humankind to schoolchildren might be validly done with the clear secular intent of enhancing the effectiveness of science instruction." Only time will tell how wide a door this opened.

———

Unsmudged, thank God, my Moon still queens the Heavens
as She ebbs and fulls, a Presence to glop at,

Her Old Man, made of grit not protein,
still visits my Austrian several

with His old detachment, and the old warnings
still have power to scare me: Hybris comes to
an ugly finish, Irreverence
is a greater oaf than Superstition.

—*W. H. Auden*

While evolutionists and creationists continue to spar over the content of science education in public schools, a third perspective is emerging. Some evolutionists of strong spiritual bent want the facts of biological and cosmic evolution to be taught with the flourish of religious revelation. This movement from science into myth is not, however, welcomed by all scientists. After all, scientists struggled for several centuries to make credible the pursuit of knowledge wholly outside myth. Should scientists now be leading the charge back into myth—albeit a scientifically based myth?

Proponents of this spiritual evolutionism are not clamoring for attention from local school boards. Their books and videos, as yet, are directed at the public at large. Nevertheless, they do fault standard curricula for presenting students with a caricature of evolution so barren and so divorced from meaning that it can elicit only boredom or despair. Members of this loose confederation have widely differing views as to the details of the "New Story." They also have disparate aims: should myths now emerging from the physical and life sciences replace or simply replenish traditional religious stories?

Champions of mythic extensions of scientific knowledge are beginning to have a presence in scientific gatherings. At the 1993 meeting of the American Association for the Advancement of Science, I watched scientists and theologians join in a symposium titled "Scientific Resources for a Global Religious Myth." Session organizer Loyal Rue, a professor of religion and philosophy at Luther College in Iowa, asked the participants, "Are there some story lines embedded in the vocabularies and perspectives of the sciences that might inform a compelling vision of human origins, human nature, and human fulfillment? We are asking the sciences this question: If a mythmaker came to you requesting some promising material for a global story, what would you suggest?"

Eric Chaisson (author of The Life Era*) and Brian Swimme (whose work appears later in this chapter) took up the challenge, offering the audience a sample of science-based mythmaking at its best. One of my other favorite extensionists from the field of biology, Jeffrey Wicken, was not at the meeting. He has yet to produce a book for a popular audience, but he has published several lovely essays that give a sense of his worldview and his brand of scientific mythmaking. Wicken, a biochemist, is one of the scientists studying the role of thermodynamic self-organization* *in evolution. He is also on the editorial advisory board of* Zygon: Journal of Religion and Science *(as are Edward O. Wilson and Eric Chaisson). The extract here is drawn from "The Cosmic Breath: Reflections on the Thermodynamics of Creation." Wicken's essay appeared in the December 1984 issue of* Zygon. *Copyright 1984 by the Joint Publication Board of* Zygon, *it is reprinted courtesy of the author and Blackwell Publishers.*

The Cosmic Breath

Jeffrey Wicken

The Platonic cosmos inherited by the Christian West was ill-disposed to an understanding of natural creation. In a world of fixed forms, processes that were random or adventitious could never generate anything of value. Since the good was not linked to natural process, the Platonic dualism of matter and form provided the ontological ground for a leitmotif of conquest and denial in Western ethico-religious thought, where the disorderly, the evil, might be banished from the cosmos by righteous living.

Alan Watts remarked that "by and large Western culture is a celebration of the illusion that good may exist without evil, light without darkness, pleasure without pain." Watts goes too far in this assessment, but nevertheless captures a measure of truth. A self-organizing cosmos into which physical and existential ingredients fit coherently requires polarities rather than dualities. The modern conception of the entropy principle has this polar character; its dissipative operation is the means by which the creative potencies of material nature are explored in time.

The concept of entropy is quite commensurable with the Western view of time as irreversible and cosmically significant. Yet the most instructive images on the indissoluble polarity of creation and destruction appear in the Eastern religions, where time is not at all the medium of human destiny and where good and evil are strictly matters of local human perspective. In Eastern views, the emphasis shifts from a transcendent God organizing nature from the outside to an immanent God whose own nature is inseparable from his creation. In Taoism there is no 'God' at all, but a rhythmic balance of polar forces, of which good and evil are but transient and parochial manifestations. Lao Tzu writes: "When everyone recognizes goodness as good, there is already evil. 'To be' and 'not to be' arise mutually."

The title of this essay is taken from a Hindu conception of creation, which seems in many ways a useful metaphor for the operation of the entropy principle in the universe's evolutionary differentiation. In Hindu religion, nature is conceived as the elaborate and changing mist of a great creative breath or *paramatman*. In mythical treatments this breath flows rhythmically through two aspects of the Godhead: Brahma the creator and Shiva the destroyer. There is no dualism in this distinction—only a recognition of polar aspects to a unitary act.

The breathing in of Shiva is quite literally an inspiration, a gathering up of the world's various manifestations for yet new expressions in a subsequent expiration of Brahma. Creation and destruction are thus but kaleidoscopic shifts in the world of *maya* or appearances. In this, as in Taoism, a certain balance of what we parochially perceive as good and evil is maintained; but finally, there is only one Supreme Self to whom all being must be referred.

Entropic Exhalation and Hindu Myth

The evolutionary differentiation of nature is a kind of entropic exhalation, a breathing out of new forms under the dissipative demands of the second law and the constraints imposed by nature's various forces. Indeed, entropic dissipation can be regarded as the breath of Brahma, the means by which the diversity of nature is made manifest.

In a cyclic universe of alternating periods of explosion and implosion, one can regard the creative potency of the Brahman breath as deriving from a prior Shivan inhalation of nature's material expressions into thermodynamic potential. We might indeed hope for these one-to-one correspondences between mythical representations and the understandings of science, but we by no means require them for the power of the divine breath metaphor to be appreciated. Even if the universe lacks sufficient mass for gravitational closure, one can still productively regard its expansion and differentiation as a great cosmic exhalation deriving from a perhaps unknowable ground of being. The driving force of the cosmic exhalation is the universe's physical expansion, which establishes a thermodynamic disequilibrium for the entropic breath to enter, carving out potential microstates for the evolutionary spread of matter-energy. The texture of the entropic breath is constituted by the cosmos's particular evolutionary unfolding.

The strategy of the cosmic breath is structuring-through-dissipation. The word dissipation refers simply to the entropic process of converting thermodynamic potential to either heat or configurational disorder. Nature's forces are, with the exception of electromagnetic repulsion, associative ones. Dissipation is the driving force of the integrative tendency in matter; it is a principle of potency, a "breath" acting to bring the possible into the actual.

Integrative processes thus become means for dissipating thermodynamic potential. Organisms not only exist through dissipation; they exist *for the reason* of dissipation as well, coming into being as stable pathways of entropy production, the breath of their own dissipation at one with the breath of the cosmic flow. Simon Black expressed it this way: "It has been tacitly assumed that life processes have a primary objective: the propagation of organisms. Energy is perceived as subserving this objective, and it seems strange that it should be thus diverted from its degradative tendency to a constructure role. But if the picture is inverted and the primary objective of life processes is assumed to be energy dissipation, and the evolution of organisms is perceived as arising secondarily to subserve this primary function, then life appears consonant with all other natural processes, which perform a predominantly dissipative role."

Again, one sees the polarity of creation-dissipation. The importance of Shiva, the fire-destroyer, to the creative process is amply celebrated in Indian mythology. I quote here from Arthur Peacocke:

Within a fiery circle representing the action of material energy and matter in nature, Shiva Nataraja (as 'he' is called in this aspect of his being) dances the dance of wisdom and enlightenment to maintain the life of the cosmos and to give release to those who seek him. In one of his two right hands he holds a drum which touches the fiery circle and by its pulsating waves of sound awakens matter to join in the dance; his other right hand is raised in a protecting gesture of hope—"do not fear"—while one of the left hands brings destructive fire to the encircling nature, and this fire, by erasing old forms, allows new ones to be evoked in the dance.

To be sure, there is an aspect here of the divine juggler, the master illusionist operating in the world of *maya*. But there is recognition too of the dynamics of creation, which requires the prior loosening of matter from its old bonds. The metaphorical power of the image lies in its recognition of the polarity of creation involving dissipation as its dynamic partner in process. In the entropic world, creation and dissipation do not come in cycles, but are conjoined aspects of a unitary process, a single breath.

This presiding image of fire as a power of transformation is essential to understanding the creative power of the entropy principle. Inside Shiva's ring the fire is creative; outside it is degradative. The entropy principle has just this kind of Janus-faced character. One sees this fundamental understanding in Heraclitus's dictum: "all things are an exchange for fire, and fire for all things."

It is essential to hold this polar unity of creation and entropic dissipation carefully in mind because the Western picture, drawn from an historical confluence of Plato, medieval Christianity and classical thermodynamics, has so colored our feelings about the entropy principle that creative processes, especially those involving the self-organization of life, seem out of synchrony with the ordinary flow of things. Indeed, the Platonic perspective has historically inspired vitalistic solutions to the problem of life, special integrative principles that run counter to, or at least independent of, the second law. But when dualisms are cemented into polarities, such notions are obviated.

Drawing from Western Ethical Traditions

We have seen that the Hindu idea of a cosmic breath concretely expresses the polar unity of the creative process, and provides an apt metaphor for thermodynamic causation. What the Eastern views fail to provide is an

adequate sense of time as the cumulative carrier of newness, in ourselves and in nature, in which progress might be a meaningful concept and in which human beings might be essential players on its evolutionary stage rather than transient configurations of the cosmic dice. Vishnu inveighs on the truth-seeker Arjuna to revel in the world of *maya* for all its worth, while leaving the grand determination of events to himself: "I am the eternal world-destroying time, manifested here for the destruction of these people. Even without thee, none of these warriors, arrayed here in the hostile armies, shall live. Therefore, do thou rise and acquire glory." This is not at all a teleological world in the Platonic tradition, where ethical decisions can be of real importance in the course of things. It is here that Western views, with their special insistence on the freedom of the individual to render good or cosmic havoc if he or she chooses, seem to contain the missing dimension of our experience of time as a vehicle for self-determined change—our power of choice to make a real difference in the course of things.

There is a deep sense in which we, in our self-determining humanity, are participants in the cosmic breath—in its sentience and its moral rightness or wrongness. Self-transcendence seems a natural tendency for sentient life, since individual viability or fitness is embedded in the network of societal-ecological relationships that constitutes the world's thermodynamic flows. Life is a thermodynamic phenomenon in a much more significant way than it is a gravitational or electromagnetic one. While these forces are involved in structuring matter in special configurations, life is nevertheless not of their nature, because time and irreversibility are not of their nature. The second law conditions the ethical realm by investing life with a dimension of temporal extension and contingency, establishing thereby the plan of goal-oriented behavior for attaining stability in time. In this, living forms, especially humans, determine nature's trajectory from the array of options available to it as a physical system.

Ethics involves a reaching out of the self to a higher-order context; in this sense, it represents an extension, made available by consciousness, of thermodynamic selection for stable patterns of entropy production. In this expansion of interest, ethics, in its broadest sense of value-laden precepts and activities, goes beyond the preservation of individuals and organizational types to an opening of nature to its highest possibilities. Thus consciousness, whose proximate adaptive payoff is biological survival, is preadapted to something far more significant—the evolution of the cosmos toward ever-greater sentient manifestation, by its capacity to substitute decision and volition for the blind necessity of survival and reproduction.

The cosmic breath, in which nature is conceived as both effect and participant, holds an image of God as the ultimate polarity of immanence and transcendence—always more than the sum of its manifestations, always the deeper ground of being from which individual beings derive their existences. Although we are denied a simplistic natural theology whereby the order of nature can be regarded as directly revelatory of God's plan, the tying of life to an ontologically complex nature by thermodynamic flows opens us up to far richer and more subtle theological possibilities, which would seem to offer renewed promise of making science and religion partners, rather than antagonists, in investing the human condition with meaning.

———

I can hear the sizzle of newborn stars,
and know that anything of meaning, of fierce
magic is emerging here. I am witness
to flexible eternity, the evolving past,
and I know we will live forever,

as dust or breath in the face of stars,
in the shifting pattern of winds.

—*Joy Harjo*

———

The Universe Story

Brian Swimme

I was presenting some ideas on the new cosmology at a conference in Chicago when suddenly a woman charged out of the audience, upset, eyes flashing as if Athena herself had decided to confront me: "I want you to explain to me why my son isn't taught this in high school. You say scientists have thrown out the materialistic world view. Then why should my son suffer it at all?"

A good question. And it concerns more than our high schools. I used to wonder something similar when I taught mathematics and physics at the university level. I was supposed to introduce students to the universe, the *universe,* but I was not to speak of meaning. Doesn't that seem like a strange assignment?

Brian Swimme is a specialist in mathematical cosmology and director of the Center for the Story of the Universe, California Institute of Integral Studies, San Francisco. Swimme's essay "The Cosmic Creation Story" appears in The Reenchantment of Science, *edited by David Ray Griffin (copyright 1988 State University of New York). The extract that follows is reprinted by permission of the author and the State University of New York Press.*

Cosmic storytelling is the central political and economic act of our time. By telling our cosmic creation story, we inaugurate a new era of human and planetary health.

A cosmic creation story is that which satisfies the questions asked by humans fresh out of the womb. As soon as they get here and learn the language, children ask the cosmic questions. Where did everything come from? What is going on? Why are you doing such and such anyway?

By cosmic creation story I also mean to indicate those accounts of the universe we told each other around the evening fires for most of the last 50,000 years. These cosmic stories were the way the first humans chose to initiate and install their young into the universe. The rituals, the traditions, the taboos, the ethics, the techniques, the customs, and the values all had as their core a cosmic story. The story provided the central cohesion for each society. *Story* in this sense is 'world-interpretation'—a likely account of the development and nature and value of things in this world.

Why Story?

Why should story be fundamental? Because without storytelling, we lose contact with our basic realities in this world. We lose contact because only through story can we fully recognize our existence in time.

Eskimo Story Teller
(Rie Muñoz)

To be human is to be in a story. To forget one's story is to go insane. All the tribal peoples show an awareness of the connection between health and storytelling. The original humans will have their cosmic stories just as surely as they will have their food and drink. Our ancestors recognized that the universe, at its most basic level, is story. Each creature is story. Humans enter this world and awaken to a simple truth: "We must find our story within this great epic of being."

What about our situation today? Do we tell stories? We most certainly do, even if we do not call them stories. In our century's textbooks—for use in grade schools and high schools—we learn that it all began with impoverished primitives, marched through the technical inventions of the scientific period, and culminated— this is usually only implied, but there is never much doubt—in the United States of America, its political freedom and, most of all, in its superior modes of production. Throughout our educational experiences, we were drawn into an emotional bonding with our society, so that it was only natural we would want to support, defend, and extend our society's values and accomplishments. Of course, this was not considered story; we were learning the facts.

Obviously, Soviets reflecting on their educational process recall a different story, one that began with the same denigration of the primal peoples, continued through a critique of bourgeois societies, and culminated in the USSR. And the French or British, reflecting on their educations, remember learning that, in fact, *they* were the important societies, for they were extending European cultural tradition, while avoiding both the superficiality of the Americans and the lugubriousness of the Soviets.

Although we told ourselves such human stories, none of us in the industrial countries taught our children cosmic stories. We focused entirely on the human world when telling our stories of value and meaning. The universe and Earth taken together were merely backdrop. The oceans were large, the species many, yes—but these immensities were just the stage for the humans.

This mistake is the fundamental mistake of our era. In a sentence, I summarize my position this way: all our disasters today are directly related to our having been raised in cultures that ignored the cosmos for an exclusive focus on the human. Our uses of land, our uses of technology, our uses of each other are flawed in many ways but due fundamentally to the same folly. We fail in so grotesque a manner because we were never initiated into the realities and values of the universe. Without the benefit of a cosmic story that provided meaning to our existence as Earthlings, we were stranded in an abstract world and left to invent nuclear weapons and chemical biocides and ruinous exploitations and waste.

Story Succumbs to Science; Science Elicits Story

How could this have happened? How could modern Western culture escape a 50,000-year-old tradition of telling cosmic stories? We discovered science. So impressed were we with this blinding light, we simply threw out the cosmic stories for the knowledge that the sciences provided. Why tell the story of the Sun as a God when we knew the sun was a locus of thermonuclear reactions? We pursued 'scientific law', relegating story and myth to the nurseries and tribes. Science gave us the real, and the best science was mathematical science. We traded myth for mathematics and, without realizing it, we entered upon an intellectual quest that had for its goal a complete escape from the shifting sands of the temporal world.

We now realize—following the work of Einstein, Hubble, and others—that ours is a universe that had a beginning in time and has been developing from 15 to 20 billion years. And every moment of this universe is new. That is, we now realize that we live not in a static Newtonian space; we live within an ongoing cosmic story. A reenchantment with the universe happens. A new love affair between humans and the universe happens.

A central desire of scientists in the future will be to explore and celebrate the enveloping Great Mystery—the story of the universe, the journey of the galaxies, the adventure of the planet Earth and all of its life forms. Scientific theories will no longer be seen simply as objective laws.

Scientific understanding will be valued as that power capable of evoking in humans a deep intimacy with reality.

I am convinced that the story of the universe that has come out of three centuries of modern scientific work will be recognized as a supreme human achievement, the scientific enterprise's central gift to humanity, a revelation having a status equal to that of the great religious revelations of the past. Suddenly, the human species as a whole has a common cosmic story. Islamic people, Hopi people, Christian people, Marxist people, and Hindu people can all agree in a basic sense on the birth of the Sun, on development of the Earth, the species of life, and human cultures. For the first time in human existence, we have a cosmic story that is not tied to one cultural tradition, or to a political ideology, but instead gathers every human group into its meanings. Certainly we must not be naive about this claim of universality. Every statement of the cosmic story will be placed in its own cultural context, and each context is, to varying degrees, expressive of political, religious, and cultural perspectives. But given that fact, we have even so broken through to a story that is panhuman; a story that is already taught and developed on every continent and within every major cultural setting.

What does this mean? Every tribe knows the central value of its cosmic story in uniting its people. The same will be true for us. We are now creating the common story which will enable *Homo sapiens* to become a cohesive community. Instead of structuring American society on its own human story, or Soviet society on its own human story, and so on, we have the opportunity to tell instead the cosmic story, and the oceanic story, and the mammalian story, so that instead of building our lives and our society's meanings around the various human stories alone, we can build our lives and societies around the Earth story.

Calling Forth the Poets

This is a good place to make my final comment on the meaning of cosmic creation story. Although with this phrase I refer in general to the account of our emergence out of the fireball and into galaxies and stars and Earth's life, I also think of the cosmic story as something that has not yet emerged. I think we will only have a common story for the human community when poets tell us the story. For until artists, poets, mystics, nature lovers tell the story—or until the poetic and mystical dimensions of humans are drawn forth in every person who sets out to tell us our story—we have only facts and theories.

Most tribal communities understand the necessity of developing storytellers—people who spend their lives learning the cosmic story and celebrating it in poetry, chant, dance, painting, music. The life of the tribe is woven around such celebrations. The telling of the story is understood both as that which installs the young and that which regenerates creation. The ritual of telling the story is understood as a cosmic event. Unless the story is sung and danced, the universe suffers from decay and fatigue. Everything depends on telling the story—the health of the people, the health of the soil, the health of the sun, the health of the soul, the health of the sky.

We need to keep the tribal perspective in mind when we examine our situation in the modern period. Instead of poets, we had one-eyed scientists and theologians. Neither of these high priests nor any of the rest of us was capable of celebrating the cosmic story. It is no wonder then that so many of us are sick and disabled, that the soils have gone bad, that the sky is covered with soot, and that the waters are filled with evils. Because we had no celebrations inaugurating us into the the universe, the whole world has become disabled.

But what will happen when the storytellers emerge? What will happen when "the primal mind" (to use Jamake Highwater's term) sings of our common origin, our stupendous journey, our immense good fortune? We will become Earthlings. We will have evoked out of the depths of the human psyche those qualities enabling our transformation from disease to health. They will sing our epic of being, and stirring up from our roots will be a vast awe, an enduring gratitude, the astonishment of communion experiences, and the realization of cosmic adventure.

We must encourage cosmic storytellers because our dominant culture is blind to their value. Is it not remarkable that we can obtain several hundred books on how to get a divorce, how to invest money, how to lose fat, and yet there is nothing available to assist those destined to sing to us the great epic of reality?

I suggest that when the artists of the cosmic story arrive, our monoindustrial assault and suicide will end and the new beginnings of the Earth will be at hand. Our situation is similar to that of the early Christians. They had nothing—nothing but a profound revelatory experience. They did nothing—nothing but wander about telling a new story. And yet the Western world entered a transformation from which it has never recovered.

So too with our moment. We have nothing compared to the massive accumulation of hate, fear, and arrogance that the intercontinental ballistic missiles, the third world debt, and the chemical toxins represent. But we are in the midst of a revelatory experience of the universe that must be compared in its magnitude with those of the great religious revelations. And we need only wander about telling this new story to ignite a transformation of humanity.

Brian Swimme has been doing just that since his book The Universe Is a Green Dragon *was published in 1984. That book is an oracular dialogue between "Youth" and "Thomas" (actually, Thomas Berry, who put an ecological slant onto the Teilhardian worldview in chapter 6). In 1992 Swimme and Berry joined to produce* The Universe Story. *There, they attempt to craft a myth of cosmogenesis, drawing from traditional stories*

*to name the characters that science itself calls forth. Tiamat is the star
whose supernova explosion provided the mineral-rich stardust that gave
birth to earth and ourselves; Aries is the first prokaryotic cell; Promethio
invents photosynthesis; Prospero learns how to respire with oxygen; Vi-
kengla is the first eukaryotic cell, Kronos the first predator. "Only now
can we see with clarity that we live not so much in a cosmos as in a
cosmogenesis, a cosmogenesis best presented in narrative; scientific in its
data, mythic in its form."*

I am memory alive
                  not just a name
but an intricate part
of this web of motion,
meaning: earth, sky, stars circling
my heart

        centrifugal.

—Joy Harjo

Gaia Theory as a Source of Evolutionary Mythmaking

Only now are we finally realizing that the Earth is a self as well. We
must embrace and cherish our dreams for the Earth. We must come to
understand that these dreams of ours do not originate in our brains
alone. We are the space where the Earth dreams.

The character Thomas speaks these words in Brian Swimme's book, The
Universe Is a Green Dragon: A Cosmic Creation Story. *Thus, the blend
of science and Teilhardian cosmology espoused by Berry and Swimme
can also be made inspirational for those inclined toward a truly Earth-
based spirituality. By itself, a shift in perspective away from the cosmos
and back to our very own planet can foster its own unique synthesis, its
own myths.*

*In the 1970s independent scientist James Lovelock developed the scien-
tific basis for a theory that the living and nonliving processes enveloping
the earth and giving rise to its dynamic texture can be viewed as a
global-scale entity. All living things, in combination with the atmosphere,
oceans, and soils (in large part created by life), form a biosphere deserv-
ing recognition in the hierarchy of life. This system, this entity expresses
an individuality that merits a name, a proper name. Lovelock (on the
recommendation of novelist William Golding) called it Gaia.*

*Biologist Laurence Levine has explored the mythic appeal and flower-
ing of new rituals brought about by the Gaia hypothesis. Gaia's appeal is
both paganistic—lending a scientific hand to renewal of a goddess tradi-*

tion now fostered by ecofeminists—and pantheistic—serving well the spiritual aspirations of some proponents of deep ecology. Levine's essay, "Gaia: Goddess and Idea" (copyright 1991 by The Humanist Institute), appeared in volume 6 of the journal Humanism Today. *An extract is reprinted courtesy of the author and the Humanist Institute.*

Laurence Levine

Gaia is the one who gives us birth
She's the air, she's the sea, she's Mother Earth
She's the creatures that crawl and swim and fly.
She's the growing grass, she's you and I.

—*from "Gaia Song"*

The chorus of children's voices sends these lyrics rising to fill the hall with their sweet innocent tones. They have just finished their responsive readings and soon the dancers will come and cast long writhing shadows before the fires. It is around Christmas time, the dark terrors of winter are still with them, and every person must help to make it through to Spring. What better way is there than to buoy their Goddess Gaia with song and dance and a rousing session on theory.

Theory? No, this is not a Druid rite, but the annual festival, "Gaia Song," given by the Commonwealth Institute of London under the partial sponsorship of IBM. And the theory the children will learn has to do with ecology, not mythology, even though it bears the name of the ancient Greek goddess for Earth.

The theory states that Earth acts as a living thing, a 'superorganism' served by its constituents—living and nonliving—in the same way that the organs of the body serve the person. The rocks, oceans and atmosphere, the microbes, people and plants, participate together in tightly integrated systems to regulate oxygen, carbon dioxide, and temperature at levels fit for life. So we are here thanks to Gaia and shall stay because she can also heal herself.

Gaia has a broad constituency. Obviously, the Goddess holds special appeal for those interested in ecological and holistic thinking. Many spiritually minded persons find her irresistible. She shines, too, for the feminist extremists, the throngs of whole-grain enthusiasts, and many other types often assembled under the "New Age" umbrella.

A publishing industry has sprung up to supply disciples with Gaian books, musings on massage, messages about endangered species, and instructions on lifestyles. Gaian organizations provide platforms for their adherents. The Liechtenstein-based Foundation for Gaia, the Gaia Institute of the Cathedral of St. John the Divine in New York City, founded by Lindisfarne members, offer seminars, lectures, and entertainment programs.

James Morton, the Dean of that institution, has found the Gaia principle useful in bringing strays back into the fold. In his hands, Gaia becomes a tool to counter the apathy, alienation and secular pursuits that compete with religion. As he explains, Gaia points to the greater whole in a religious way since Gaia does what religion means to do—to knit together again. He has commissioned Paul Winter's help in the knitting process. This composer, noted for the incorporation of nature sounds into his music, has created a Missa Gaia. Now wolf calls, whale sounds, and the voice of the loon rise with the Kyrie and resonate with the Sanctus and the Benediction among the high Episcopalian arches.

For scientists, Gaia is a launching platform into discovery and cross-disciplinary thinking. For the disenchanted it is a perch, and for those who have strayed from the fold, a source of renewed spiritual energy. Even secular humanists may feel her draw as a metaphor for transcendence without God. To be sure, few of us who have seriously dealt with the Gaia hypothesis will ever see Earth and life in the same way again.

As a long-time "stray" I can vouch for the power of ritual derived from a synthesis of Gaian science and the splendor of Episcopalian surrounds. Like tens of thousands of other New Yorkers, I attend the annual Winter Solstice celebration at the Cathedral of St. John the Divine. Inside this spectacular gothic cathedral, I swoon to the echoes of wolflike songs that flow from Paul Winter's saxophone and that herald the rise of a luminous Earth balloon, spinning slowly upward, a jewel in a dark and infinite space.

——

The earth has dreamed me to stand
on the rise of this highway to admire who
she has become.

—Joy Harjo

*The Love Embrace
of the Universe*
(Frida Kahlo)

Acknowledgments

I thank all living authors for promptly reviewing my selections from their works and for allowing me the latitude to construct workable and flowing essays—often from one or more books. Their words of encouragement and (in several instances) intercessions with publishers helped me keep the administrative burdens of this project from blotting out the joy. Special thanks go to William Provine for bringing to my attention relevant passages from Darwin's autobiography (which appear in chapter 10) and to Peter Corning for suggesting that I use *synergy* instead of *integration* in the title of chapter 4.) Richard Dawkins was instrumental in bringing to my attention the need to make transparent the extent of the restructuring of essays and to direct readers to specific page numbers in original sources: hence the concordance.

I am indebted to my fellow contributors of original works in chapter 8, Dorion Sagan and Mitchell Waldrop, who not only came through but did so in good humor and in the face of an unmovable deadline. Three anonymous reviewers for MIT Press made helpful suggestions.

I thank Jean Medawar for reviewing my selections from a book review by her late husband, Peter Medawar. I am indebted to Mary Catherine Bateson for approving passages I selected from the work of Gregory Bateson. Lynn Margulis and Crispin Tickell both expended effort in trying to find a member of the Huxley family to review my selections from the works of Julian Huxley. Léo Laporte graciously reviewed my choice of excerpts from works by the late G. G. Simpson. (Thanks to Niles Eldredge for suggesting I contact Laporte.) Marc Swetlitz provided advice on the Huxley and Simpson debate on progress. Sarah Appleton-Weber critiqued my initial construction of an essay from the book by Pierre Teilhard de Chardin. I credit Arthur Fabel for bringing to my attention pertinent passages from E. O. Wilson's latest book and for pressing me to read Mitchell Waldrop's book, *Complexity*. I am especially grateful to Carter for putting me in touch with the American Teilhard Society and for introducing me to the works of Brian Swimme and Thomas Berry. John C. Greene has my thanks for general advice and encouragement.

Cliff Matthews introduced me to the poetry of Joy Harjo. Ed Dobb directed me to a cornucopia of poetry with a science slant: *Songs from Unsung Worlds: Science in Poetry* (edited by B. B. Gordon, Birkhäuser and AAAS, 1985). I am grateful to Anthony Blake, too, for poetry advice.

Harry M. Buck and the journal *Anima* helped me track down the owner of *Monkey Reaching for the Moon*. Mark McMenamin, Michael Rampino, and Lynn Margulis supplied photostats of several figures. John Tyler Bonner gave advice for the illustration of myxobacteria. Artist Margaret La Farge authorized a reprinting of her original art. Rie Muñoz Ltd., again in this book as in the first, donated a photograph of a watercolor by Rie Muñoz. I thank Marian Vlasic for his gift of an original drawing. I am indebted to Eric Finklestein for introducing me to the drawings of trees by Paul Landacre. Barbara A. Brown and Richard Sloss at the American Museum of Natural History helped me attach names to the organisms in several of the old scientific illustrations that appear here. Finally, I thank the Photograph Library of the Metropolitan Museum of Art for friendly and efficient service and for making it possible for the public to browse through photographs of the museum's vast collection of paintings and sculpture. I am also grateful for the open-stack policy of the Bobst Library at New York University, which made it easy for me to discover century-old science books filled with superb line drawings. The interlibrary loan staff of Miller Library at Western New Mexico University provided essential help during my year away from New York.

Because I was unable to obtain financial support to help with permissions and other costs, I am grateful to the publishers that heeded my plea for leniency in permissions fees: Harvard University Press, University of Chicago Press, Yale University Press, Princeton University Press, Oxford University Press, University of Arizona Press, Sterling Lord Literistic Inc., Jason Aronson Inc., Kluwer Academic Publishers, Open Court Publishing Co., *Zygon: The Journal of Religion and Science,* the journal *Social Research,* the National Association of Biology Teachers, the Vatican Observatory, the Institute for Creation Research, and the Humanist Institute. I am also grateful for the generosity of Diane Ackerman, Annie Dillard, The Literary Estate of May Swenson, and Curtis Brown Ltd. with respect to poetry excerpts that appear here.

I extend warm thanks to James Lovelock, whose lecture at the University of Washington in 1985 compelled me to give science a second look, and to Richard Dawkins for having written the book (*Blind Watchmaker*) that inspired me to assemble this anthology series. I will be forever grateful to Lynn Margulis for helping my manuscript (*From Gaia to Selfish Genes*) find a publisher and for mentoring me in numerous ways. The MIT Press has proved to be an ideal publisher; I owe special thanks to Yasuyo Iguchi and the design department for exquisite taste, to Madeline Sunley and Ann Sochi in the acquisitions department for personal attention and wise counsel, and to director Frank Urbanowski for sharing my vision.

Finally, I thank Tyler Volk for scientific and literary advice and for emotional support throughout this long project. A partner in science, a partner in wilderness adventure, and a partner in life: were everyone as fortunate as I.

Bibliography and Further Reading

Note: For exact page numbers of the longer extracts, consult the concordance.

Chapter 1: Prophets of Progress is primarily selections from four of Julian Huxley's books, all of which portray evolution as progressive: *Essays of a Biologist* (1923, Knopf), *Evolution: The Modern Synthesis* (1943, Harper & Row), *Evolution in Action* (1953, Harper & Row), and *New Bottles for New Wine* (1957, Harper & Row). Within the text Huxley refers to a book by John B. Bury, *The Idea of Progress* (1920/1932, Macmillan), and to J. B. S. Haldane's *The Causes of Evolution* (1932, Princeton University Press).

Huxley's arguments are supplemented by those of Francisco J. Ayala, drawn from an essay in *Evolutionary Progress,* edited by Matthew H. Nitecki (1988, University of Chicago Press). An earlier version of this essay appeared in a book edited by Ayala and Theodosius Dobzhansky: *Studies in the Philosophy of Biology* (1974, University of California Press). Portions of the essay not reprinted here review attempts by various biologists to define progress (or its absence) in evolution. Nitecki's edited volume on progress is a sampling of the thoughts of contemporary evolutionary biologists: David Hull, William Provine, Michael Ruse, Robert J. Richards, John Maynard Smith, David Raup, Stephen Jay Gould, and others.

Closing the chapter are extracts from a 1992 book by Edward O. Wilson, *The Diversity of Life* (Harvard University Press). For further reading consult *The Basis of Progressive Evolution* by evolutionary biologist G. Ledyard Stebbins (1969, University of North Carolina Press). Progressivists who focus on the increase in brain size through time ("encephalization") include Dale A. Russell and H. J. Jerison. Russell's ideas are set forth in his essay "Speculations on the Evolution of Intelligence in Multicellular Organisms," which appeared in *Life in the Universe,* edited by John Billingham (1981, MIT Press). Jerison's classic book is *Evolution of the Brain and Intelligence* (1973, Academic Press).

For an overview of the idea of progress, including its manifestations in science, consult *Good Tidings: The Belief in Progress from Darwin to Marcuse* by W. Warren Wagar (1972, Indiana University Press). For a treatment of pre-Darwinian scales of reference, see Arthur O. Lovejoy's classic, *The Great Chain of Being* (1936, Harvard University Press).

Chapter 2: The Naysayers begins with selections from the writings of George Gaylord Simpson: *This View of Life* (1964, Harcourt Brace), *The Meaning of Evolution* (1949, Yale University Press), "The Nonprevalence of Humanoids" (*Science* 143:769–77), and "The Concept of Progress in Organic Evolution" (*Social Research,* spring 1974). The book by C. J. Herrick cited by Simpson is *The Evolution of Human Nature* (1956, University of Texas Press).

Excerpts from John Tyler Bonner's *Evolution of Complexity* (1989, Princeton University Press) follow Simpson's pieces. Bonner refers to a 1962 paper by Herbert A. Simon, "The Architecture of Complexity," *Proceedings of the American Philosophical Society* 106: 467–82. Next, the 1988 book *Evolutionary Progress* is drawn from once again, with brief quotations by editor Matthew H. Nitecki and contributor Stephen Jay Gould. Two more short quotations of Gould come from pages 44 and 91 of Gould's *Wonderful Life* (1989, Norton). Gould's conclusions in that book have not gone unchallenged; see John Maynard Smith's essay "Taking a Chance on Evolution" in 14 May 1992 *New York Review of Books* and Derek Briggs et al. in a 1992 issue of *Science* 256:1670–73 (with follow-up debate in 258:1816–18). Henry Gee wrote a review of the debate for *Nature* 358:456–57.

The chapter concludes with selections from *Extinction: Bad Genes or Bad Luck?* by David Raup (1991, Norton). Raup refers to a book by Clark Chapman and David Morrison: *Cosmic Catastrophes* (1989, Plenum Press). He also mentions the break-through work by Luis Alvarez and colleagues on the role of extraterrestrial impacts in mass extinctions; refer to Alvarez's autobiography (*Alvarez: Adventures of a Physicist*, 1987, Basic Books) for an anecdotal account of how he made his discoveries. The historic paper that disclosed the Alvarez et al. discovery appeared in a 1980 issue of *Science* (208:1095–1108). Michael Rampino's reinterpretation of glacial tillites as impact ejecta is in press at the journal *Geology*; a popular account appears in the January 1994 issue of *Earth*. For biological implications of Rampino's theory, see "Did Surface Temperatures Constrain Microbial Evolution?" by David Schwartzman, Mark McMenamin, and Tyler Volk. This paper appeared in the June 1993 issue of *BioScience* 43:390–93).

An important work refuting the notion of progress in evolutionary biology (too technical, however, to include in this book) is a chapter "Natural Selection, Adaptation, and Progress" by George C. Williams. It appears in his classic 1966 book, *Adaptation and Natural Selection* (Princeton University Press). Finally, philosopher of science Marc Swetlitz devotes his doctoral thesis to the debate between Huxley and Simpson: "Julian Huxley, George Gaylord Simpson, and the Idea of Progress in Twentieth Century Evolutionary Biology" is available through University Microfilms International (Ann Arbor, Michigan).

Chapter 3: Wedges, Arms Races, and the Role of Strife begins with a laudatory quotation by George Gaylord Simpson, drawn from page 268 of his *The Meaning of Evolution*. Darwin's quotation about mosquitoes comes from page 116 of his *The Voyage of the Beagle* (as it appears in the Bantam Classic reprint edition). Darwin's "wedge" quotation is drawn from his short essay "On the Variation of Organic Beings in a State of Nature," which was read at the Linnean Society in 1858 and was then copublished with Wallace's similar essay in the *Journal of the Proceedings of the Linnean Society of London, Zoology* in 1859. Excerpts from *The Origin of Species* were drawn from the sixth edition; the concordance keys to the pagination in the Modern Library edition, published by Random House.

The essay by Richard Dawkins was selected from his *The Blind Watchmaker* (1986, Norton). Mention is made of the Life/Dinner Principle, which was set forth in a 1979 paper by Richard Dawkins and John R. Krebs, "Arms Races Between and Within Species," *Proceedings of the Royal Society of London, Series B* 205:489–511. The 1973 paper by Leigh van Valen on the Red Queen hypothesis is "A New Evolutionary Law," *Evolutionary Theory* 1:1–30. The work by R. A. Fisher that Dawkins cites is the classic 1930 book, *The Genetical Theory of Natural Selection* (Dover). Malte Andersson's research on widow birds was published in 1982 in *Nature* 299:818–20. For another, and surprising, view of how evolutionary arms races proceed, consult a 1992 paper by Wolfgang Sterrer, "Prometheus and Proteus: The Creative, Unpredictable Individual in Evolution," *Evolution and Cognition* 1:101–29. Finally, for a look at the entire research program owing to the genic perspective developed by Dawkins and others, refer to the 1992 book by Helena Cronin, *The Ant and the Peacock* (Cambridge University Press).

Chapter 4: Governors, Tinkerers, and the Role of Synergy begins with a short essay pieced together from two books by Gregory Bateson: *Steps to an Ecology of Mind* (1972, Ballantine) and *Mind and Nature* (1979, Dutton). Bateson can be difficult; a good way to ease into his ideas is through *Angels Fear* (1987, Macmillan), which was assembled and coauthored by his daughter, Mary Catherine Bateson, from his notes and their conversations just prior to his death. The quotation of Alfred Russel Wallace appears in Wallace's short essay "On the Tendency of Varieties to Depart Indefinitely from the Original Type," which was copublished with a short essay by Charles Darwin in 1859 in the *Journal of the Proceedings of the Linnean Society of London, Zoology.*

The essay on evolutionary "tinkering" by François Jacob appeared first in 1977 in *Science* 196:1161–66. It was later revised for a special lecture series at the University of Washington and then appeared as a chapter in *The Possible and the Actual* (1982, Pantheon).

The essay by Lynn Margulis and Mark McMenamin was drawn from "Marriage of Convenience," which appeared in the September 1990 issue of *The Sciences.* For more on symbiosis see Margulis's classic book, *Symbiosis in Cell Evolution* (1993/1981, Freeman). Margulis is coeditor (with René Fester) of *Symbiosis as a Source of Evolutionary Innovation* (1991, MIT Press). She is also coauthor (with Dorion Sagan) of a popular account of the evolutionary story, *Microcosmos* (1986, Simon and Schuster). Finally, she and McMenamin served as editors for the English-language version of *Concepts of Symbiogenesis,* by Liya Nikolaevna Khakhina (1992, Yale University Press). For an in-depth look at the Garden of Ediacara hypothesis, see *The Emergence of Animals: The Cambrian Breakthrough* by Mark McMenamin and Dianna Schulte McMenamin (1990, Columbia University Press).

The chapter closes with excerpts from *The Synergism Hypothesis: A Theory of Progressive Evolution* by Peter A. Corning (1983, McGraw-Hill). He refers to a book by Edward O. Wilson, *Sociobiology: The New Synthesis* (1975, Harvard University Press), and to a two-part paper by William Hamilton, "The Genetical Evolution of Social Behavior" (1964, *Journal of Theoretical Biology* 7:1–52). Corning also mentions a 1971 paper by Robert Trivers, "The Evolution of Reciprocal Altruism" (*Quarterly Review of Biology* 46:35–57). For more on the interweaving of competition and cooperation, see a review by John Rennie, "Living Together," which appeared in the January 1992 issue of *Scientific American.* See also Ashley Montagu's *Darwin: Competition and Cooperation* (1952, Henry Schuman) and the classic by Peter Kropotkin, *Mutual Aid* (1987/1902, Freedom Press).

Chapter 5: Ratchets, Uroboros, and the Role of Initiative begins with two quotations of Jacob Bronowski. The first is from page 308 of *The Ascent of Man* (1973, Little, Brown), the second from page 76 of *Nature and Knowledge* (1969, Oregon State System of Higher Education, Eugene). The Bronowski essay, "New Concepts in the Evolution of Complexity: Stratifed Stability and Unbounded Plans," was published several times: *Boston Studies in the Philosophy of Science* 8 (1970, Kluwer), *Synthese* 21 (1970), and *Zygon* 5 (1970). The concordance paginates to the *Zygon* publication.

My own summary of the work in self-organization deriving from a thermodynamic perspective refers to work by Manfred Eigen. His latest book is *Steps Toward Life* (1992, Oxford University Press). Carl Jung's description of the uroboros comes from his *Psychology and Alchemy* (1908, Princeton University Press). The two paragraphs by Stuart Kauffman are drawn from pages 26, xiii, and 173 of his *Origins of Order: Self-Organization and Selection in Evolution* (1993, Oxford University Press).

Rod Swenson's idea about the role of life in maximizing entropy production can be found in his forthcoming book, *Spontaneous Order, Evolution, and Natural Law* (Erlbaum), and in his 1989 "Emergent Attractors and the Law of Maximum Entropy Production: Foundations for a General Theory of Evolution," *Systems Research* 6:186–97. The paper by Bruce Weber et al. is "Evolution in Thermodynamic Perspective: An Ecological Approach," which appeared in 1989 in *Biology and Philosophy* 4:373–405. With Jeffrey S. Wicken, Stanley Salthe, David Depew, and others as coauthors, this paper provides a fine sampling of the ideas floating around in the self-organization research community. Its list of references, moreover, is a good representation of the field. Bruce Weber and David Depew coauthored *Darwinism Evolving* (1994, MIT Press).

For further reading consult *Evolution, Themodynamics, and Information* by Jeffrey S. Wicken (1987, Oxford University Press). Many other technical books have been produced on various aspects of self-organization, which Hermann Haken and Buckminster Fuller (independently) named "synergetics." About fifteen years ago each published a book by that title. One book that is both technical and popular is *The Self-Organizing Universe* by Erich Jantsch (1980, Pergamon).

Alister Hardy's essay is drawn from *The Living Stream: Evolution and Man* (1965, Harper & Row). This book provides a delightful introduction to the basics of biological evolution. It came to my attention as a citation in a 1967 book by Arthur Koestler, *The Ghost in the Machine* (Hutchinson; Viking Penguin). In the chapter "Progress by Initiative" Koestler presents and elaborates on Hardy's views. Philosopher of science Karl Popper cites Hardy, too, in his *Unended Quest: An Intellectual Autobiography* (1976, Open Court), from which the closing essay in this chapter is drawn. Portions of Popper's selections are also drawn from *Of Clouds and Clocks: An Approach to the Problem of Rationality and the Freedom of Man*. This slim volume presents Popper's Compton Memorial Lecture, delivered at and published by Washington University (St. Louis, Missouri). See also Popper's 1990 book, *A World of Propensities* (Thoemmes, Bristol). A much-cited book on the role of initiative in evolution, which introduces the concept of "organic selection," is *Development and Evolution* by James Mark Baldwin (1902, Macmillan).

Chapter 6: Cosmic Visions begins with a lengthy excerpt from Pierre Teilhard de Chardin's classic book, *The Phenomenon of Man*. An English translation was published in 1959 by William Collins and Harper & Row; it is now available through HarperCollins. Teilhard scholar Sarah Appleton-Weber is working on a new translation for HarperCollins.

Following Teilhard is a short quotation by Thomas Berry, president emeritus of the American Teilhard Association, who modernizes Teilhard's vision to mesh with ecological concerns. The quotation is drawn from the 1982 pamphlet *Teilhard in the Ecological Age* (Anima Books). Excerpts from Julian Huxley's book *New Bottles for New Wine* (1957, Harper & Row) close the chapter. Huxley mentions an essay written by his grandfather, Thomas Henry Huxley; the essay "Agnosticism" was published in 1889.

Teilhard's vision gave birth to a vast outpouring of writings. A sympathetic reworking and critique of Teilhard was produced by the evolutionary biologist Edward O. Dodson, *The Phenomenon of Man Revisited* (1984, Columbia University Press). For other contemporary viewpoints that use Teilhard as a foundation, consult the American Teilhard Association (40 Hillside Lane, Syosset, NY 11791).

Chapter 7: Banishing Cosmic Meaning begins with two often-cited critiques of Teilhard's cosmic vision. Both appeared as book reviews; the brief quotation drawn from G. G. Simpson's friendly critique was published in the April 1960 issue of *Scientific*

American (pp. 201–6), while Peter Medawar's scathing review appeared in a 1961 issue of the journal *Mind* 70:99–106, reprinted in Medawar's 1984 book, *Pluto's Republic* (Oxford University Press). For a list of other critiques, consult the final section in E. O. Dodson's previously cited book.

Francis Crick's praise of Monod's book appeared in a 1976 issue of *Nature* (262:429–30). The bulk of this chapter comprises excerpts from Jacques Monod's *Chance and Necessity* (1971, Knopf). It is the definitive evolutionary account of the universe from an existentialist perspective.

Chapter 8: Beyond the Binary begins with brief denouncements of Monod's worldview by Mary Midgley and Stanley Salthe. Midgley's critique is drawn from her 1985 book, *Evolution as a Religion* (Methuen). Salthe's comes from page 175 of his essay in *Evolutionary Biology at the Crossroads,* edited by Max K. Hecht (1989, Queens College Press). For a fuller expression of his views and his critique of Monod, see Salthe's 1992 "Science as the Basis for a New Mythological Understanding," *Uroboros* 2:25–43.

The critique by Theodosius Dobzhansky is an amalgam of his writings. The first, third, and last paragraphs are drawn from his "Essay on Religion, Death, and Evolutionary Adaptation," which appeared in a 1966 issue of *Zygon* 1:317–31. Paragraph 2 comes from his essay "Two Contrasting World Views," which appears in *Beyond Chance and Necessity,* edited by John Lewis (1974, Humanities Press). Paragraph 4 is drawn from page 115 of Dobzhansky's 1967 book, *The Biology of Ultimate Concern* (New American Library). Another excellent essay is Dobzhansky's "Chance and Creativity in Evolution," which appeared in *Studies in the Philosophy of Biology: Reduction and Related Problems,* edited by Francisco Ayala and Theodosius Dobzhansky (1974, University of California Press).

Dorion Sagan introduces the ideas of Vladimir Vernadsky. Unfortunately, the limited amount of Vernadsky's work that has been published in or translated into English is rather dense. Refer to his "The Biosphere and the Noösphere," published posthumously in 1945 in the journal *American Scientist* 33:1–12. The Vernadskian view is, nevertheless, well represented in serviceable English translations of two books by Russian scientists: *Traces of Bygone Biospheres* by A. V. Lapo (1982) and *Evolution of the Biosphere* by M. M. Kamshilov (1976). Both books were published by Mir, Moscow; the publishing arm of Biosphere 2 in Oracle, Arizona, may be the best place to find them in the United States.

Dorion Sagan's own book is *Biospheres: Reproducing Planet Earth* (1990, McGraw-Hill). For more on the strange nature of biological individuality see Sagan's "Metametazoa: Biology and Multiplicity," pages 362–85 in *Incorporations (Zone 6; Fragments for a History of the Human Body),* edited by Jonathan Crary and Sanford Kwinter (1992, Zone). On his "Narcissus" theme, refer to Sagan's "What Narcissus Saw: The Oceanic 'I/Eye,'" pages 247–66 in *Speculation, The Reality Club,* edited by John Brockman (1990, Prentice-Hall).

For an introduction to the Gaia hypothesis, see the entry "Gaia Hypothesis" by Dorion Sagan and Lynn Margulis in *McGraw-Hill Yearbook of Science and Technology, 1993.* Also, read the first three chapters in *From Gaia to Selfish Genes,* edited by Connie Barlow (1991, MIT Press). The originator of the Gaia hypothesis, James Lovelock, has produced three books on the subject: *Gaia: A New Look at Life on Earth* (1979, Oxford University Press), *The Ages of Gaia* (1988, Norton), and *Healing Gaia* (1991, Crown). A 1992 technical review paper is (my own) "Gaia and Evolutionary Biology" by Connie Barlow and Tyler Volk (*BioScience* 42:686–93). A compilation of technical papers drawn from the first scientific conference on Gaia is *Scientists on Gaia,* edited by S. H. Schneider and P. E. Boston (1991, MIT Press).

The book that launched the selfish genic worldview and research program is *The Selfish Gene* by Richard Dawkins (1976, Oxford University Press). My own edited volume, *From Gaia to Selfish Genes,* contains excerpts from this world-shaking and eloquent little book and from important papers that followed. Dawkins's second book, *The Extended Phenotype* (1982, Freeman), takes the selfish gene concept several leaps further.

The "triad of new books" on complexity are *Complexity: The Emerging Science at the Edge of Order and Chaos* by M. Mitchell Waldrop (1992, Simon and Schuster); *Complexity: Life at the Edge of Chaos* by Roger Lewin (1992, Macmillan); and *Artificial Life: The Quest for a New Creation* by Steven Levy (1992, Pantheon). Also recommended is Stuart Kauffman's *The Origins of Order: Self-Organization and Selection in Evolution* (1993, Oxford University Press). Chris Langton organized the first Artificial Life conference and edited the first volume of proceedings, *Artificial Life* (1989, Addison-Wesley). His own essay is an excellent introduction to the principles underlying complexity, as well as to the titled field he named and helped launch. Langton is first of four coeditors of *Artificial Life II* (1992, Addison-Wesley). See also the book series by Gordon and Breach Science Publishers, The World Futures General Evolution Studies; forthcoming is *Chaos and the Evolving Ecological Universe* by S. J. Goerner.

Chapter 9: Evolution as Religion begins with excerpts from *On Human Nature* by E. O. Wilson (1978, Harvard University Press). Julian Huxley follows, with excerpts drawn from his timeless *Religion Without Revelation* (1957, Harper & Row). The passage by Alister Hardy is drawn from Hardy's *Darwin and the Spirit of Man* (1984, William Collins). The chapter concludes with a few paragraphs by historian of science John C. Greene, drawn from his *Science, Ideology, and World View: Essays in the History of Evolutionary Ideas* (1981, University of California Press). See also his 1990 essay, "The Interaction of Science and World View in Sir Julian Huxley's Evolutionary Biology," *J. History of Biology* 23:39–55.

For more on this topic consult the books and essays by Mary Midgley and Theodosius Dobzhansky cited for the previous chapter. Ralph Wendell Burhoe (whom Wilson mentions in his essay) produced "Religion's Role in Human Evolution: The Missing Link Between Ape-Man's Selfish Genes and Civilized Altruism" (1979, *Zygon* 14:135–62). I once spent several days perusing all the back issues of *Zygon: Journal of Religion and Science* and found that it contained a wealth of fine essays on many issues dealt with in this anthology. Arthur Peacocke comments on the ideas of many of the people who appear in this anthology in a chapter in *Darwinism and Divinity,* edited by John Durant (1985, Blackwell).

Chapter 10: Responding to Creationism begins with a statement by the pragmatist philosopher Charles Sanders Pierce; it appeared in Pierce's 1893 essay "Evolutionary Love" (*Monist* 3:176–200), reprinted in *Darwinism and the American Intellectual,* edited by R. J. Wilson (1967, Dorsey Press). I particularly recommend the essay in that book by Louis Agassiz, whose nineteenth-century views on this subject are very similar to that of today's creationists. The classic book expressing the majority creationist worldview in pre-Darwinian times is *Natural Theology* by Archdeacon William Paley (1802). In the 1830s, eight contributors wrote in a similar vein in *The Bridgewater Treatises.*

The selections by Henry M. Morris come from his 1985/1982 book, *What Is Creation Science?* cowritten with Gary E. Parker. It is available through the Institute for Creation Research (ICR) (San Diego, California). For those who would like to learn about the scientific arguments in favor of creation (and against evolution), this book is the place to begin. Another ICR publication, *Evolution: The Challenge of the Fossil Record,* by Duane T. Gish (1985) is also a good introduction. Morris asserts that humanism is a religion; for a history of this issue, see "How Religious Is Secular Humanism?" by Leo Pfeffer in the September 1988 issue of *The Humanist.*

The biographical note about Morris is drawn from "Creationism in Twentieth-Century America" by Ronald L. Numbers, which appeared in a 1982 issue of *Science,* 218:538–44, and was reprinted with other useful essays in the June 1987 issue of *Zygon.* Numbers has since produced a book, *The Creationists: The Evolution of Scientific Creationism* (1992, Knopf). Next, a passage is drawn from pages 82 and 63 of "Remarkable Birth of Planet Earth," a 1972 fundamentalist tract by Henry Morris that is quoted on page 56 of *Science and Creationism,* edited by Ashley Montagu (1984, Oxford University Press.)

The section on the liberal religious response begins with Charles Darwin; initial passages are drawn from *The Origin of Species* (Modern Library, sixth edition), which are followed by excerpts from the unexpurgated version of *The Autobiography of Charles Darwin,* edited by Darwin's grand-daughter, Nora Barlow (1958, Harcourt Brace). Passages from Theodosius Dobzhansky's "Nothing in Biology Makes Sense Except in the Light of Evolution" appeared in *American Biology Teacher* (March 1973, pp. 125–29).

The section on the separatist response begins with a passage from the October 1981 "Address by Pope John Paul II to the Pontifical Academy of Science." It can be found in *But Is It Science?* edited by Michael Ruse (1988, Prometheus). Many useful essays and court opinions can be found in that book. Selections from a more recent statement by the pope were drawn from *John Paul II on Science and Religion: Reflections on the New View from Rome,* edited by Robert J. Russell, William R. Stoeger, and George V. Coyne (1990, Vatican Observatory); this book also contains commentary on the pope's statement by a variety of contributors. The pope's statement (and essays by others) can be found as well in *Physics, Philosophy, and Theology: A Common Quest for Understanding,* also edited by Russell, Stoeger, and Coyne, and published in 1988 by the Vatican Observatory.

The aphorism by James Madison comes from his 1784 essay, "Memorial in Remonstrance." The brief quotation by philosopher of science Philip Kitcher is drawn from pages 175–76 of his book *Abusing Science: The Case Against Creation* (1982, MIT Press). Ashley Montagu speaks next, from pages 9 and 14 of the book he edited, *Science and Creationism* (1984, Oxford University Press). He is followed by evolutionary biologists Niles Eldredge, whose remark is drawn from a 1981 letter to the editor of *Science* 212:737, and Stephen Jay Gould (*Science and Creationism,* pp. 124–25). Karl Popper's viewpoint is drawn from pages 341–42 of his "Natural Selection and the Emergence of Mind" (1978, *Dialectica* 3:339–55). John C. Greene's statement comes from pages 196–97 of his *Science, Ideology, and World View* (1981, University of California Press).

The section on the hardliner response begins with a passage by Francis Bacon drawn from his statement number 65 in *Novum Organum* (1620). Next comes philosopher Mary Midgley from page 12 of her *Evolution as a Religion* (1985, Methuen), followed by George Gaylord Simpson from *This View of Life* (1964, Harcourt Brace). Concluding this section are excerpts from writings of two contemporary evolutionary biologists: William Provine's essay draws from pages 69–71 of his "Progress in Evolution and Meaning in Life" in *Evolutionary Progress,* edited by Matthew H. Nitecki (1988, University of Chicago Press); Richard Dawkins's quip comes from page 316 of his *Blind Watchmaker* (1986, Norton).

The section on the conciliatory response begins with a quotation by Alfred North Whitehead from page 186 of his *Science in the Modern World* (1925, Macmillan; reprinted 1967 by Free Press). Next come excerpts from Martin Eger's "A Tale of Two Controversies" in the September 1988 issue of *Zygon* 23:291–325. Following are several quotations by John Stuart Mill (drawn from his *On Liberty,* 1859) and a quotation by Thomas Henry Huxley (from page 229 of his *Darwiniana,* 1894). The section closes with contemporary comments by Dorothy Rosenthal (from a 1987 issue of *Science Education* 71:187–88), Kenneth W. Kemp (from a 1988 issue of *American Biology Teacher* 50:76–81), Michael Scriven (from the winter 1985 issue of *National Forum,* pp. 10–11), and Richard D. Alexander (from pages 108–9 in *Evolution Versus Creationism: The Public Education Controversy,* edited by J. Peter Zetterberg, 1983, Oryx Press). Unlike many similar anthologies, Zetterberg's book presents extracts from creationist tracts in addition to a fine selection of proevolution arguments and judicial decisions.

Extracts from the 1987 Supreme Court decision in *Edwards v. Aguillard* can be found in *Supreme Court Reporter* 107:2573–2607. The concordance keys to this reference. The chapter closes with an extract from *Trial and Error: The American Controversy over Creation and Evolution* by Edward J. Larson (1989, Oxford University Press). Of all books cited here, this in my view is the single best (indeed, pleasurable) way to gain an understanding of the cultural forces that have incited the several waves of antievolution and creationist sentiment in the United States. Another helpful and highly recommended book for the broader context is *Issues in Science and Religion* by Ian G. Barbour (1966, Prentice-Hall).

Chapter 11: Science into Myth begins with an extract from "The Cosmic Breath: Reflections on the Thermodynamics of Creation" by Jeffrey Wicken (published in the December 1984 issue of *Zygon* 19:487–505). Brian Swimme follows with extracts from *The Universe Is a Green Dragon* (1984, Bear and Company) and "The Cosmic Creation Story," which appeared in *The Reenchantment of Science,* edited by David Ray Griffin (1988, State University of New York Press). Swimme's latest—and inspirational—book is *The Universe Story* (1992, HarperCollins). Thomas Berry is coauthor; Berry's earlier book is *Dream of the Earth* (1988, Sierra Club Books). Swimme has a twelve-part video series, *Canticle to the Cosmos,* available through Tides Foundation, Livermore, California.

The chapter concludes with a short passage from an essay by biologist Laurence Levine, "Gaia: Goddess and Idea," which appeared in 1991 in *Humanism Today* (volume 6). Many books and essays have spun off the science of the Gaia hypothesis and into the realm of speculative philosophy. For a blend of both, consult *Gaia: The Human Journey from Chaos to Cosmos* by Elisabet Sahtouris (1989, Simon and Schuster).

Eric Chaisson propounds his own version of cosmic evolution in *The Life Era* (1987, Norton). Consult, too, his 1988 essay, "Our Cosmic Heritage," in *Zygon* 23:469–79. Louise B. Young presents a similar story in *The Unfinished Universe* (1986, Simon and Schuster).

Goethe, Emerson, Wordsworth, Blake, Carlyle, Dante, Sir Thomas Brown, Shelley, and the rest of the assembly of immortal spirits—they jostle each other on your shelves, each waiting only to be picked up to introduce you to his own unique and intense experience of reality.

—*Julian Huxley*

Concordance

Page citations from the original books and articles are grouped by section head and listed in order of appearance. One or more citations are provided for each paragraph of text. Slashes demarcate the paragraphs. Consult the bibliography or text for complete bibliographic information.

1 Prophets of Progress

Julian Huxley *Essays of a Biologist* (B); *Evolution: The Modern Synthesis* (M); *Evolution in Action* (A); *New Bottles for New Wine* (N)

Progress: Myth or Science? N-20, A-125 / A-125 / N-20 / N-42 / N-21

Critics of "Higher" and "Lower" A-87 / B-10 / M-556 / B-11 / B-12, M-567, B-12 / B-12

Countering the Critics A-85 / N-291 / N-46, N-47 / N-47 / N-47 / B-13

Succession of Dominant Types M-558 / M-559 / M-559 / M-559 / M-560 / M-560

Efficiency, Control, and Independence M-561 / M-562 / M-562 / M-562 / M-562 / M-563 / M-563, M-570 / M-564, M-570

The Pageant of Life N-45 / N-291, B-25 / N-252 / A-89 / B-27

Defining Evolutionary Progress B-30, M-565 / N-91

Is Progress Inevitable? N-27 / A-128, B-35

Accounting for Cruelty B-38, B-40 / B-40 / A-87, A-88 / A-88 / A-89

Excesses of Anthropomorphism M-565 / M-565 / M-565 / M-566

Human Purpose in Evolution M-576 / M-576 / M-577 / M-577 / M-577 / M-578 / M-578

Human Society in the Cosmic Process A-163 / A-166 / A-150 / N-293 / N-40

Francisco Ayala *Evolutionary Progress*

An Alternative View of Progress 88 / 95 / 91 / 91 / 92 / 92 / 92 / 92 / 93 / 93 / 93 / 93 / 94 / 94 / 94 / 95

Edward O. Wilson *The Diversity of Life*

Biological Diversity as the Scale of Progress 186 / 187 / 187 / 187 / 187 / 188 / 188 / 190 / 190 / 190, 344 / 187 / 351

2 The Naysayers

George Gaylord Simpson *This View of Life* (V); *The Meaning of Evolution* (M); "The Concept of Progress in Organic Evolution" (C); "The Nonprevalence of Humanoids" (H)

[no section head] V-37

The Idea of Progress M-239 / M-240 / M-241 / M-241

Evolution and the Fossil Record H-773 / H-773 / C-37 / H-773

Countering Julian Huxley C-46, 47 / C-47, 48 / C-48 / C-48

Man Is Not the Measure C-49 / M-250 / M-251 / M-251

Progress in Perception M-257 / M-258 / M-259 / M-259

Ad Hoc Improvements C-50 / C-51 / M-242 / M-260

Synergism in Social Life 79 / 79 / 80 / 80 / 88 / 88
Egoistic Co-operation 103, 117 / 84 / 85 / 103 / 104 / 104, 108
Synergism and Progressive Evolution 93 / 93 / 95 / 96 / 96 / 104 / 96
Beyond the Binary 256 / 256 / 256 / 257 / 257

5 Ratchets, Uroboros, and the Role of Initiative

Jacob Bronowski   *Ascent of Man* (A); *Nature and Knowledge* (N); (in) *Zygon* 5 (Z)
Stratified Stability and Unbounded Plans A-308 / N-76 / Z-18 / Z-18 / Z-19 / Z-30
The Barb of Evolution Z-30 / Z-31 / Z-31 / Z-31 / Z-31 / Z-31 / Z-32 / Z-32
Thermodynamics and the Arrow of Time Z-32 / Z-32 / Z-33 / Z-33 / Z-33 / Z-33 / Z-34 / Z-34

Alister Hardy   *The Living Stream*
The Role of Initiative 10, 34 / 34 / 154 / 154 / 155, 160
Behaviour as a Selective Force 161 / 170 / 171, 170, 177 / 171 / 171 / 171 / 171 / 172 / 172 / 172
The Example of Darwin's Finches 174 / 174 / 174 / 175
The Role of Behaviour in Adaptive Radiation 192 / 198 / 199 / 200 / 200 / 207

Karl Popper   *Unended Quest* (U); *Of Clouds and Clocks* (C)
Changes in Preferences, Then Skills, Then Anatomy U-173 / U-173 / U-174 / U-174 / U-175 / U-176
The Critical Attitude and Cultural Evolution C-23 / C-24 / C-26 / C-26 / C-26 / C-26, 19, 21 / C-27 / C-27 / C-30

6 Cosmic Visions

Pierre Teilhard de Chardin   *The Phenomenon of Man*
Body and Soul 53 / 53 / 53 / 62 / 62 / 63 / 64 / 65
A Line of Progress 134 / 142 / 143 / 142
Threshold of Reflection 143 / 144, 146 / 146 / 165 / 168 / 169 / 165 / 160
A New Age 178 / 178, 180 / 183 / 183
Birth of the Noosphere 181 / 181 / 182 / 182 / 215 / 220 / 220 / 221 / 219
The Source of Our Disquiet 227 / 227 / 227 / 228 / 228 / 228 / 229
The Road Ahead 233 / 233 / 231 / 231, 232
Appearances of a Setback 257 / 257 / 255 / 255
The Personal and the Universal 261 / 261 / 261, 262 / 257, 259 / 260 / 263
Spiritual Renovation of Earth 244 / 244 / 245 / 242 / 243, 267 / 267
Love 264 / 264 / 265 / 265 / 266 / 266
Summary 290, 36 / 252 / 253 / 283, 284 / 284, 30

Julian Huxley   *New Bottles for New Wine*
A Secular Version of Cosmic Evolution 13 / 13 / 13, 14 / 100
The Sequence of Evolution 101, 102 / 101 / 101 / 102 / 102 / 103 / 103, 121
The Need for New Beliefs 287 / 289 / 110 / 293
An Evolutionary Ethic 294 / 295 / 295 / 300
Restoring Our Unity with Nature 301 / 302 / 122
The Scientific Spirit 304 / 305 / 305
On Population, Eugenics, and Art 306 / 306 / 307
Personal Reflections 310 / 311 / 311 / 311 / 312 / 312

7 Banishing Cosmic Meaning

Jacques Monod   *Chance and Necessity*
Chance and Necessity 43 / 44 / xiii
The Animist Covenant 29 / 30 / 31 / 31 / 31 / 31 / 32 / 33 / 33

9   Evolution as Religion

10   Responding to Creationism

11 Science into Myth

Jeffrey Wicken "The Cosmic Breath"
The Cosmic Breath 492 / 492 / 492 / 493 / 493
Entropic Exhalation and Hindu Myth 494 / 494 / 494 / 495 / 496 / 496 / 496 / 496
Drawing from Western Ethical Traditions 498 / 499 / 500 / 503
Brian Swimme "The Cosmic Creation Story" (C); *The Universe is a Green Dragon* (G); *The Universe Story* (U)
The Universe Story G-17 / G-17 / C-47 / C-48 / C-48
Why Story? C-48 / C-48 / C-48 / C-48 / C-49 / C-49
Story Succumbs to Science; Science Elicits Story C-49 / C-50, C-51 / C-51 / C-51, C-52 / C-52
Calling Forth the Poets C-52 / C-53 / C-53 / C-53 / C-53 / C-53 / C-53 / U-229

Sources of Illustrations

Preface

Slug. In Alfred S. Romer, *The Procession of Life* (World Publishing Co., 1968).

Chapter 1

Lingula. In S. F. Harmer and A. E. Shipley, *The Cambridge Natural History*, vol. 3 (Macmillan, 1895).

Trilobites, eurypterids, and horseshoe crabs. In Ernst Haeckel, *Kunstformen der Natur* (1904; reproduced in *Art Forms in Nature*, Dover, 1974.)

Anatomy of air supply system. Plate IX in H. G. Bronn, *Tier-Reichs*, vol. 5 (Akademische Verlagsgesellschaft, 1925).

Duck on Rock in Snowy Moonlit Night, by Shibata Zeshin (1806–1891). Album leaf, lacquer on paper. The Metropolitan Museum of Art, The Howard Mansfield Collection, Rogers Fund, 1936. (36.100.106)

Monkey Reaching for the Moon in the Water, by Hakuin Ekaku (1685–1768). Hanging scroll, ink on paper. Courtesy of Gotoh Museum, Tokyo.

A relative of *Paramecium*. Plate VIII in Friedrich Stein, *Infusionsthiere*, vol. 1 (Verlag Von Wilhelm Engelmann, 1859).

Venus flytrap and sundew. In Anton Joseph Kerner, *The Natural History of Plants*, 1894–95.

Maple Leaves, by Shibata Zeshin (1806–1891). Lacquer on paper. The Metropolitan Museum of Art, The Howard Mansfield Collection, Rogers Fund, 1936. (36.100.134)

Ediacaran fossils. In Mark A. S. McMenamin and Dianna L. Schulte McMenamin, *The Emergence of Animals* (Columbia University Press, 1990). Courtesy of Mark McMenamin.

Ten Butterflies, by Shibata Zeshin (1806–1891). Lacquer on paper. The Metropolitan Museum of Art, The Howard Mansfield Collection, Rogers Fund, 1936. (36.100.111)

Chapter 2

Bristlecone Pine, by Paul Landacre. In Donald Culross Peattie, *A Natural History of Western Trees* (Houghton Mifflin, 1950). Illustrations copyright 1953 and renewed 1981 by Paul Landacre. Reprinted by permission of Houghton Mifflin Co.

Fer-de-lance (a pit viper). Plate III in *Royal Society of London Philosophical Transactions*, volume 94, 1804.

Fruiting bodies of various myxobacteria. Plate IIVI in *Botanical Gazette*, volume 37, 1904.

Volvox. In Alfred S. Romer, *The Procession of Life* (World Publishing Co., 1968).

Anatomy of a sponge. In Alfred S. Romer, *The Procession of Life.* (World Publishing Co., 1968).

Gliders. In Wilhelm Haacke, *Die Schöpfung der Tierwelt* (1893).

Ammonite fossils. In Alfred E. Brehm, *Merveilles de la natur* (1893).

Foraminifera. In Ernst Haeckel, *Kunstformen der Natur.* (1904; reproduced in *Art Forms in Nature,* Dover, 1974.)

Permian impact crater, proposed by Michael Rampino. Courtesy of Michael Rampino, New York University.

Foxtail Pine, by Paul Landacre. In Donald Culross Peattie, *A Natural History of Western Trees* (Houghton Mifflin, 1950). Illustrations copyright 1953 and renewed 1981 by Paul Landacre. Reprinted by permission of Houghton Mifflin Co.

Chapter 3

Mylodon darwinii. In Charles Darwin, *Journal of Researches* (1890; reproduced in Alan Moorehead, *Darwin and the Beagle,* Harper & Row, 1969).

Invasion of locusts. In Alfred E. Brehm, *Merveilles de la natur* (1893).

Dryptosaurus, by Charles Knight, 1897. Negative no. 335199. Courtesy Department of Library Science, American Museum of Natural History.

Water beetle. In John Richardson, William S. Dallas, and T. Spencer Cobbold, *The Museum of Natural History,* vol. 4 (Virtue and Yorston, 1877).

Spider, bird, butterfly. In Jim Harter, *Animals: 1419 Copyright-Free Illustrations of Mammals, Birds, Fish, Insects, etc.* (Dover Publications, 1979).

Arsinotherium and Pterodon, by Charles Knight, 1907. Negative no. 35832. Courtesy Department of Library Services, American Museum of Natural Historv.

Trunks of trees, by Agnes Miller Parker. In H. E. Bates, *Through the Woods* (Macmillan, 1936).

Bird of paradise. In Jim Harter, *Animals: 1419 Copyright-Free Illustrations of Mammals, Birds, Fish, Insects, etc.* (Dover Publications, 1979).

Chapter 4

Ladder to the Moon, by Georgia O'Keeffe, 1958. Oil on canvas. Collection of Emily Fisher Landau.

Duck-billed platypus. In M. Charles D'Orbigny, *Dictionnaire universel d'histoire naturelle* (1849; reproduced in Oliver Goldsmith, *The Illustrated History of the Natural World,* Random House, 1990).

Flounders. In Alfred E. Brehm, *Merveilles de la natur* (1893).

Queen conch, *Strombus gigas.* Plate 3 of Strombus in volume 2 of Jean-Charles Chenu, *Illustrations conchylioques* (1843).

Dance of the chromosomes. In Edmund B. Wilson, *The Cell in Development and Heredity* (Macmillan, 1928).

Mixotricha paradoxa, by Christie Lyons. By permission of Lynn Margulis.

Prairie dog town. In John George Wood, *Homes Without Hands* (1866/1892).

Eagle Claw and Bean Necklace, by Georgia O'Keeffe, 1934. Charcoal. The Museum of Modern Art, New York. Given anonymously (by exchange).

Chapter 5

Sting of bee. In Philip H. Gosse, *Evenings at the Microscope* (1860).

Tapeworm. In Brayton Howard Ransome, "The Taeniod Cestodes of North American Birds," *U.S. National Museum Bulletin,* vol. 69 (1909).

Bur of *Xanthium spinosum.* After detail of figure 196 in Leroy G. Holm et al., *The World's Worst Weeds* (University Press of Hawaii, 1977).

Mantis. In Wilhelm Haacke, *Die Schöpfung der Tierwelt* (1893).

Dragon devouring its tail. In Michael Maier, *Atlanta Fugiens* (1618; facsimile edition 1964).

Uroboros. In Eleazar, *Uraltes Chymiches Werk* (1760; reproduced in C. G. Jung, *Psychology and Alchemy,* Princeton University Press, 1980).

Flying fish. In Alfred E. Brehm, *Merveilles de la natur* (1893).

Blue tit and bottle, by Margaret La Farge. In John T. Bonner, *The Evolution of Culture in Animals* (Princeton University Press, 1980). By permission of Margaret La Farge.

Heads of Galapagos finches. In Charles Darwin, *Journal of Researches* (1897; reproduced in Alan Moorehead, *Darwin and the Beagle,* Harper & Row, 1969).

Galapagos tortoises and finches. In Alfred E. Brehm, *Merveilles de la natur* (1893).

Marsupial mammals. In Wilhelm Haacke, *Die Schöpfung der Tierwelt* (1893).

Chapter 6

The Reunion of the Soul and the Body, by William Blake, 1813 (from *The Grave Etching*). The Metropolitan Museum of Art, Harris Brisbane Dick Fund, 1917. (17.3.2400)

Dante Astray in the Dusky Wood, by Gustave Doré, 1861–1868. Illustration for Dante's *The Divine Comedy.*

The Empyrean, by Gustave Doré, 1861–1868. Illustration for Dante's *The Divine Comedy.*

The Ascent, by Gustave Doré, 1861–1868. Illustration for Dante's *The Divine Comedy.*

Lammergeier. In John George Wood, *Bible Animals* (Longmans, 1869).

Chapter 7

Native American rock art. In Campbell Grant, *Rock Art of the American Indian* (Promontory Press, 1967).

Elément méchanique I, by Fernand Leger, 1924. Oil on canvas. Smith College Museum of Art, Northampton, Massachusetts.

Fish on mangrove roots. In Alfred E. Brehm, *Merveilles de la natur* (1893).

The Scream, by Edvard Munch (1863–1944). Lithograph. The Metropolitan Museum of Art, Bequest of Scofield Thayer, 1982. (1984.1203.1)

Chapter 8

Untitled I, by Marian Vlasic, 1992. Ink on paper. Courtesy of Marian Vlasic, Zaturecka 16, Bratislava 83107, Slovak Republic.

Untitled II, by Marian Vlasic, 1992. Ink on paper. Courtesy of Marian Vlasic, Zaturecka 16, Bratislava 83107, Slovak Republic.

Chapter 9

God Strikes Down the Rebellious Angels, by Gustave Doré, 1865. Illustration for Milton's *Paradise Lost.*

The Eagle, by Gustave Doré, 1861–1868. Illustration for Dante's *The Divine Comedy.*

Terrestrial Paradise, by Gustave Doré, 1861–1868. Illustration for Dante's *The Divine Comedy.*

Virgil and Dante, by Gustave Doré, 1861–1868. Illustration for Dante's *The Divine Comedy.*

Chapter 10

The Creation of Fishes, by Gustave Doré, 1865. Illustration for Milton's *Paradise Lost.*

Clam Shell, by Georgia O'Keeffe (1887–1986). Oil on canvas. The Metropolitan Museum of Art, Alfred Stieglitz Collection, 1962. (62.258)

Archaeopteryx. In Wilhelm Haacke, *Die Schöpfung der Tierwelt* (1893).

The Seventh Evening in Eden, by Gustave Doré, 1865. Illustration for Milton's *Paradise Lost.*

Chapter 11

Shiva Nataraja. Metropolitan Museum of Art. Bronze (South India, circa A.D. 1000). The Metropolitan Museum of Art, Harris Brisbane Dick Fund, 1964. (64.251)

Eskimo Story Teller, by Rie Muñoz, 1973. Watercolor. Copyright 1973 Rie Muñoz Ltd., Juneau, Alaska. Courtesy Rie Muñoz, Ltd.

The Love Embrace of the Universe, by Frida Kahlo, 1949. Oil on masonite. Jacques and Natasha Gelman Collection.

Acknowledgments

Sea slug. In Alfred S. Romer, *The Procession of Life* (World Publishing Co., 1968).

Sources of Poetry

Preface

Diane Ackerman. From "Lady Faustus" in *Jaguar of Sweet Laughter: New and Selected Poems* by Diane Ackerman, copyright 1991 by Diane Ackerman (Random House); originally in *Lady Faustus,* 1976 (Morrow). Reprinted courtesy of Diane Ackerman.

Chapter 1

Walt Whitman. From "Walt Whitman" in *Leaves of Grass* by Walt Whitman, 1855.

W. H. Auden. "Progress?" in *Collected Poems* by W. H. Auden, copyright 1974 by the Estate of W. H. Auden. Reprinted by permission of Random House, Inc. and courtesy of Curtis Brown Ltd.

Julian Huxley. From page 2 of *Essays of a Biologist* by Julian Huxley, 1923. Reprinted by permission of the Huxley Estate.

Chapter 2

Annie Dillard. From "Bivouac" in *Tickets for a Prayer Wheel* by Annie Dillard, copyright 1974 by Annie Dillard (Harper & Row). Reprinted courtesy of Annie Dillard.

May Swenson. From "The Universe" in *New and Selected Things Taking Place* by May Swenson, copyright 1963 by May Swenson (renewed 1991). Reprinted courtesy of the Literary Estate of May Swenson.

Diane Ackerman. From "Mars" in *Jaguar of Sweet Laughter: New and Selected Poems* by Diane Ackerman, copyright 1991 by Diane Ackerman (Random House); originally in *The Planets: A Cosmic Pastoral,* 1976. Reprinted courtesy of Diane Ackerman.

Chapter 3

Annie Dillard. From a prose paragraph on page 242 of *Pilgrim at Tinker Creek* by Annie Dillard, copyright 1974 by Annie Dillard (Harper's Magazine Press). Line ends and use as poetry approved by the author. Reprinted courtesy of Annie Dillard.

Alfred, Lord Tennyson. From sections LV and LVI of "In Memoriam," 1833.

Chapter 4

Rainer Maria Rilke. From "Just as the Winged Energy of Delight, " February 1924.

*Carrol B. Fleming. From "Boundaries," copyright 1983 by Carrol B. Fleming. Published in *Songs from Unsung Worlds: Science in Poetry,* edited by Bonnie Bilyeu Gordon, copyright 1985, American Association for the Advancement of Science.

Walt Whitman. From "Song of Myself " in *Leaves of Grass* by Walt Whitman, 1855.

Chuang-Tzu. From "The Floods of Autumn," no. 5, sec. 10, bk 17, pt. 2 of *The Writings of Chuang-Tzu.*

Chapter 5

William James. From a prose sentence on page 144 of *The Pluralistic Universe* by William James, 1909 (reprinted 1977 by Harvard University Press). Line ends determined by the editor.

Edwin Stuart Russell. Attributed by Alister Hardy (page 45 of *The Living Stream,* 1965) to page 61 of *The Study of Living Things* by E. S. Russell (undated).

Robinson Jeffers. From "De Rerum Virtute" in *Selected Poems* by Robinson Jeffers, copyright 1954 by Robinson Jeffers. Reprinted by permission of Random House, Inc.

Edward Young. From "The Consolation," Night IX in *Night Thoughts* by Edward Young, 1856.

Chapter 6

Walt Whitman. From "Carol of Occupations" in *Leaves of Grass,* 1855.

Walt Whitman. From "Walt Whitman" in *Leaves of Grass,* 1855.

Chapter 7

Voltaire. From "Egyptian Monuments," section XX in *The Philosophy of History,* 1764.

William Wordsworth. From "Ode: Intimations of Immortality," 1804.

Chapter 8

*Robert Frazier. "The Supremacy of Bacteria," copyright 1980 by Robert Frazier. Published in *Songs from Unsung Worlds: Science in Poetry,* edited by Bonnie Bilyeu Gordon, copyright 1985, American Association for the Advancement of Science.

William James. From a prose paragraph on page 143 of *Pluralistic Universe,* 1909 (reprinted 1977 by Harvard University Press). Line ends determined by the editor.

Chapter 9

Alfred, Lord Tennyson. From "The World Is Too Much With Us," 1804.

Walt Whitman. From "Starting from Paumanok" in *Leaves of Grass,* 1860.

Alfred North Whitehead. From a prose paragraph on the last page of chapter 12 of *Science and the Modern World,* 1925 (Macmillan); reprinted 1967 by Free Press. Line ends determined by the editor.

George Bernard Shaw. From a prose paragraph in *Back to Methuselah: A Metabiological Pentateuch,* copyright 1921 by George Bernard Shaw (Brentano's). Line ends determined by the editor.

Chapter 10

Genesis 1:20–21.

Diane Ackerman. From "St. Augustine Contemplating the Bust of Einstein" in *Wife of Light* by Diane Ackerman, copyright 1978 by Diane Ackerman (Morrow). Rearranged by the editor, with approval of the author. Reprinted courtesy of Diane Ackerman.

Zenophanes.

Robinson Jeffers. From "Thebaid" in *The Selected Poetry of Robinson Jeffers* by Robinson Jeffers, copyright 1937 and renewed 1965 by Donnan Jeffers and Garth Jeffers. Reprinted by permission of Random House, Inc.

*Heather McHugh. From "A Physics" in *American Poetry Review,* copyright 1984 by Heather McHugh. Also published in *Songs from Unsung Worlds: Science in Poetry,* edited by Bonnie Bilyeu Gordon, copyright 1985, American Association for the Advancement of Science.

*Laura Fargas. From "Natural History" in *Songs from Unsung Worlds: Science in Poetry,* edited by Bonnie Bilyeu Gordon, copyright 1985, American Association for the Advancement of Science.

W. H. Auden. From "Moon Landing" in *Collected Poems* by W. H. Auden, copyright 1969 by W. H. Auden. Reprinted by permission of Random House, Inc.

Chapter 11

Joy Harjo. From page 56 of *Secrets from the Center of the World* by Joy Harjo and Stephen Strom (volume 17 of Sun Tracks: An American Indian Literary Series), copyright 1989 by the Arizona Board of Regents. Line ends determined by the editor, with the approval of the author (with slight revisions in text by the author). Reprinted courtesy of Joy Harjo and University of Arizona Press.

Joy Harjo. From "Skeleton of Winter" in *She Had Some Horses* by Joy Harjo, 1983 (Thunder's Mouth Press). Reprinted by permission of the author and Thunder's Mouth Press.

Joy Harjo. From page 50 of *Secrets of the Center of the World* by Joy Harjo and Stephen Strom (volume 17 of Sun Tracks: An American Indian Literary Series), copyright 1989 by the Arizona Board of Regents. Line ends determined by the editor, with the approval of the author. Reprinted courtesy of Joy Harjo and University of Arizona Press.

*I have been unable to trace through the publisher the authors of these poems, who retained the copyrights to their works published or reprinted in *Songs from Unsung Worlds: Science in Poetry,* edited by Bonnie Bilyeu Gordon, copyright 1985, American Association for the Advancement of Science.

Index

Modern synthesis, 3, 213
Mol, Hans J., 224
Mollusks, eyes of, 99–100
Monod, Jacques
  as author, 180–198
  biography of, 180–181
  criticism of, 201–202
  mentioned, 97, 207–208, 213, 214,
    217, 218, 226
Montagu, Ashley (as author), 266
Morgan, C. Lloyd, 130, 145
Morley, Lord, 231, 232
Morris, Henry
  as author, 251–258
  biography of, 252
Morton, James Parks, 299–300
Multicellularity, origins of, 47, 115
Mutations, 33, 122, 185–188
Mythopoeic drive, 193–194, 224, 226
Myths
  creation, 201–202, 224, 273, 288–290,
    293
  scientifically based, 226–230, 287–300
Myxobacteria, 42–43, 47

Natural selection. *See also* Evolution
  Darwin coins term, 75–77
  discussed, 11, 15, 33–34, 122, 167,
    185, 188, 191, 202
  inadequacy of, 126–127, 129, 132, 134,
    213
  survival of the fittest, 65
Nerve net, 13, 23
Nervous system, 13, 23–24, 41, 45, 101,
  150. *See also* Mental processes
New Age philosophy, 256–257, 299
Newton, 222, 237, 256
Newtonian worldview, 128
Nietzsche, 222
Nitecki, Matthew (as author), 49
Nucleotides, 184–185
Nucleus, origin of, 108
Numbers, Ronald L. (as author), 252
Noosphere, 145, 153, 157–159, 163,
  206, 210, 217

Objectivity, principle of, 183, 194–197,
  207, 213. *See also* Critical attitude or
  stance
Omega point, 159, 206, 217, 245
Organelles, origin of, 103–109

Organic selection, 130, 134, 136
Origin
  of life, 25, 126
  of universe, 288–290, 295
Origin of Species
  about, 64–65
  excerpts from, 66–77, 258–260
  mentioned, 239, 251, 260, 268, 276
Orthogenetic trends, 138–139. *See also*
  Evolution, direction
Ostracoderms, 8
Oxygen, origin of, 27, 104
Ozone, 27

Paley, William, 260
Paramecium, 22
Parasites, 6, 11, 16, 104–105, 211–212,
  215
Peacocke, Arthur, 290
Philosophy, western, 183
Pierce, Charles Sanders (as author), 251
Pit viper, 37–38
Planaria, 23
Plants, 22–23, 28, 44, 67–74
Plato, 183, 241, 288, 290, 291
Poetry, importance of, 239, 296–297
Pope John Paul II (as author), 263–265
Popper, Karl
  as author, 138–142, 268
  biography of, 140
  mentioned, 138, 217, 244
Population problem, 173, 237
Population thinking, 31, 213, 215
Pragmatism, 251
Predation, origin of, 47
Prigogine, Ilya, 257
Process theology, 222, 224
Progress, evolutionary
  arguments against, 5–7, 16, 20–21, 32–
    39, 49–57, 183
  arms races as cause, 79, 82–83
  definitions of, 7–11, 13–14, 18, 21, 24,
    25–26, 28, 40
  other discussion of, 3–5, 15, 18, 140,
    145–150, 157, 161, 166, 202, 245,
    247, 291
  stratified stability as cause, 121–125
  synergy as cause, 115
Prokaryotes. *See* Bacteria
Protists. *See* Protoctists
Protoctists, 22, 42, 105, 107, 109, 115